INDUSTRIAL ROBOTS

THEIR ROLE IN MANUFACTURING INDUSTRY

ORGANISATION FOR ECONOMIC CO-OPERATION AND DEVELOPMENT

Pursuant to article 1 of the Convention signed in Paris on 14th December, 1960, and which came into force on 30th September, 1961, the Organisation for Economic Co-operation and Development (OECD) shall promote policies designed:

- to achieve the highest sustainable economic growth and employment and a rising standard of living in Member countries, while maintaining financial stability, and thus to contribute to the development of the world economy;
- to contribute to sound economic expansion in Member as well as non-member countries in the process of economic development; and
- to contribute to the expansion of world trade on a multilateral, non-discriminatory basis in accordance with international obligations.

The Signatories of the Convention on the OECD are Austria, Belgium, Canada, Denmark, France, the Federal Republic of Germany, Greece, Iceland, Ireland, Italy, Luxembourg, the Netherlands, Norway, Portugal, Spain, Sweden, Switzerland, Turkey, the United Kingdom and the United States. The following countries acceded subsequently to this Convention (the dates are those on which the instruments of accession were deposited): Japan (28th April, 1964), Finland (28th January, 1969), Australia (7th June, 1971) and New Zealand (29th May, 1973).

The Socialist Federal Republic of Yugoslavia takes part in certain work of the OECD (agreement of 28th October, 1961).

Publié en français sous le titre:

ROBOTS INDUSTRIELS
LEUR RÔLE DANS L'INDUSTRIE MANUFACTURIÈRE

This study, prepared by the Secretariat of the Industry Division, was undertaken as part of the Industry Committee's on-going work on productivity in manufacturing industry. The Industry Committee, at its 50th Session in March 1983, considered and approved the report, and agreed to recommend its publication in view of the growing interest in industrial robots, their role in manufacturing industry and their expected impact on productivity. The report was derestricted by the Council of the OECD on 26th May, 1983.

Also available

TELECOMMUNICATIONS. Pressures and Policies for Change (April 1983)
(93 83 02 1) ISBN 92-64-12428-4   152 pages   £6.90   US$14.00   F69.00

ICCP (Information, Computer and Communications Policy):
No. 7 MICRO-ELECTRONICS, ROBOTICS AND JOBS (May 1983)
(93 82 02 1) ISBN 92-64-12384-9   266 pages   £12.50   US$25.00   F125.00

INDUSTRY IN TRANSITION. Experiences of the 70s and Prospects for the 80s (to be published)

TABLE OF CONTENTS

Chapter I

BASIC CONCEPTS AND DEFINITIONS

1. INTRODUCTION

The manufacturing industries of the OECD Member countries are, in the present economic situation, being adversely affected through increasing cost pressures, low productivity growth and stagnant demand. Several large industrial sectors are facing severe adjustment pressure as a result of this weak demand situation and from increasing pressure of international competition.

Nevertheless, while faced with low profit levels and cash flow problems, many manufacturing industries are increasingly investing in new automated manufacturing equipment in order to increase productivity and thus their international competitiveness. Based on developments of semiconductor technology and on new machining technology and concepts, the increasing automation of capital equipment is leading to significant changes in concepts of manufacturing and in the evolution of manufacturing equipment. The rapid development of techniques of automatic control in manufacturing processes, the integration of computers in real time directly to machines, and the integration of machines through industrial robots, will lead to the application of these new manufacturing concepts to a wide range of industries. Thus it can be expected that, during the next decade, in addition to productivity increases, and changes in the structure of production, the organisation, level and structure of employment will also change.

This study examines the implications of industrial robots on the manufacturing industries of OECD Member countries. Although the central focus of the study is on robot applications, the other aspects of automation are important. Robotics is only one facet, even though it is a major one, of computer-aided manufacturing and the trend towards increasing automation in manufacturing. It must also be kept in mind that the introduction of robots has been part of the more incremental changes in automation which have occurred in the production process over the last few decades.

2. AUTOMATED MANUFACTURING EQUIPMENT:

A Description

The concept of automation refers to the ability of a machine to perform a given sequence of tasks and meet certain specifications automatically, that is, without human intervention. In addition, the concept involves the ability of the machine to control its performance enabling the system to monitor and adjust its performance through a feedback system to ensure that the given specifications are met. Automation is thus a generic title covering mechanisation, electrification, use of electronics and the substitution of power for human efforts.

The linkage between mechanised operations and computers, which forms the basis of automation and the ability of feedback control, has been in effect since the 1940s. In recent years the use of automated manufacturing processes and equipment has accelerated in the manufacturing industries of the OECD Member countries. Although the spread of such equipment has been spurred on by miniaturisation, in particular by the application of microelectronics, it has been the manufacturing industries, particularly the mechanical engineering sector, which have been in the forefront of developments of automation relevant to industrial manufacturing. The trend of present developments has expanded the scope of mechanisation and advanced techniques in handling materials and products and of assembling components.

Computer-aided Manufacturing

Computer-aided manufacturing (CAM) forms an integral part of the developments in the automation of manufacturing equipment. This concept, which dates from the early 1950s, derives from the increasing integration of computer control in the functioning of machine tools and other manufacturing equipment. The term CAM is wide-ranging, covering elements such as:

  i) Numerically controlled machine tools (NC), computer numerical control (CNC), and direct numerical control (DNC). Numerical control is being extended into various other forms of production equipment.
  ii) automated manufacturing systems.
  iii) automated inspection systems - these have a wide range of applications.
  iv) automated warehousing - including inventory control.
  v) industrial planning and control.
  vi) industrial robots.

CAM is thus concerned with the application of automated capital equipment into the production process. The concept is often linked with Computer-aided design (CAD) which encompasses the use of computers in product design, product analysis and other engineering analysis. The efficiency of CAD lies in its linkage to machines capable of receiving data directly and adjusting their performance to carry out the required functions.

The quantification and definition of some of the elements within the area covered by CAM is difficult because it is wide-ranging and not a discrete sector as normally defined in the nomenclature of industrial activity. Some data exist, but these are often based on differing definitions and too broad in their coverage. In this regard, it is worth noting the views of the Working Party on Automation of the Economic Commission for Europe. It concluded that in order to examine the impact of automation in different industrial branches ...

"the statistical basis required for making an economic efficiency analysis does not exist on an international level, neither at the enterprise level nor the industrial sector level. One of the main reasons for this is the multipurpose aspect of equipment related to automation ..."(1)

This is in particular the case for CAD systems whereas for other individual elements under CAM (e.g. NC, CNC and DNC machine tools and industrial robots) an adequate data base exists(2).

The essential difference of some of the computer-aided manufacturing elements compared to special-purpose automated machines is that the latter are inflexible [therefore labelled "hard automation"], whereas CAM technology tends to provide flexibility because it is reprogrammable [therefore labelled "soft" programmable automation](3). This difference means that special-purpose

---

1. ECE Working Party on Automation, ECE/Automat/11, Statistical Framework for Assessing the Economic Efficiency of Automation in Industry, 1979, page 1. The document outlines the aim of the Working Party of the ECE to undertake "a statistical framework as a methodological approach for assessing the development and diffusion of automation and its economic efficiency in the main industrial branches of the economies of the ECE countries".
2. As concerns industrial robots, data bases are being developed fairly rapidly, particularly as a result of work by national standards organisations, engineering and manufacturing associations.
3. Automation in discontinous production is much more difficult as compared to process industries where production is continuous or semi-continuous.

machinery, because it is inflexible and expensive to modify in order to incorporate product changes, is suitable for high-volume production runs used in the process industries. Flexibility in some CAM systems means that they are suitable for small/medium volume production and where there is a need to incorporate frequent product changes. A comparison of CAD/CAM and process control is summarised in Figure 1.

The flexibility provided by the newer technology has meant that over time much of the so-called "hard automation" will be able to incorporate more flexibility. Thus the line of differentiation between automated capital equipment used in semi-continuous production and equipment used in batch production is becoming less rigorous. At present, industrial production is characterised on the one hand by high volume methods, emphasizing standardization and using very dedicated machinery. On the other hand production is based on small scale production methods producing a variety of non-standardized parts. The merging of flexibility provided with the less dedicated machinery in batch production techniques with the productivity resulting from mass production techniques should have important consequences in manufacturing over the next two decades.

Increased flexibility has also allowed greater integration of "hard" and "soft" automated machinery. Many of the elements that come within the concept of computer-aided manufacturing can thus be integrated together into production "systems". Thus new concepts such as Automated Batch Manufacturing Systems (ABMS)(4) and Flexible Manufacturing Systems (FMS)(5) have evolved particularly, but not exclusively, in the metal forming and mechanical engineering industries. This ability of the various elements within computer-aided manufacturing to be integrated into production systems will have important consequences, particularly as such systems form the foundation of unmanned manufacturing plants.

----

4. Defined as an electronically controlled network of flexible and dedicated machine tools linked by material handling equipment to convert stop and go batch manufacturing into continous or nearly continous processing.
5. Flexible manufacturing systems have as their components usually NC machine tools, for the production of parts, automated carrier and warehousing (storage) equipment, assembly robots and a central co-ordinating computer. FMS thus constitute the integration of automated manufacturing equipment. Care should be taken in interpreting statistics on the growth of FM systems which are often cited in conjunction with statistics on the growth in sales of NC tools and robots, and could thus involve an element of double-counting.

Figure 1

SOME SIMILARITIES AND DIFFERENCES BETWEEN
CAD/CAM AND PROCESS CONTROL

Attribute	Process Control	CAD/CAM
Computer implemented	Yes	Yes
Real time system	Yes	Yes
Embedded system	Yes	Yes
Sensor based inputs	Main source for most information in system (Pressure, temperature flow, etc.)	Minor portion of information in system. Mostly events, timing, etc.
Input of human origin	Minor portion of information in system (set points, etc.)	Major source of information. (Design configurations, production status, order status, information)
Expanding data base	No	Yes
Process control	Major purpose is feedback or feed forward control in classic sense. Major process units included within these loops, process gains and dynamics important	Orientation is more towards the mere handling, timing, release etc. of large volumes of information.
Output interfaces	Set point stations, valves, etc.	Plotters, machine tools, wiring machines, flame cutters, robotic units, automatic test equipment.
Predominant user industries	Chemical, petroleum steel, pulp and paper	Discrete parts manufacturing, (transportation equipment, machinery, etc.)
Socio-economic impact	Modest	Much larger
Main period of pioneering	1960-1975	1975-1990

Source: J. Scrimgeour, Engineering Journal (Canada), August 1981, page 8.

11

New automated capital equipment and developments in CAM technology have obvious implications for productivity changes in manufacturing and the structure of industrial production. However, this impact is often difficult to assess because:

    i) much of this equipment is being applied horizontally across industries, which disguises its real impact. Applications such as control, testing and inspection equipment, transfer equipment, automated warehousing and distribution, lend themselves to virtually every form of manufaturing activity. In addition, some applications, while important in terms of the particular workpost in which they are applied, have a marginal impact in terms of the total production process of the industry itself.

    ii) many of the concepts are too wide-ranging for assessment (automated manufacturing systems, industrial planning and control);

    iii) some of the elements under CAM have an important, but a more specific sectoral impact (NC machine tools, NC lathes, etc.);

    iv) many of the elements have only an indirect impact on major manufacturing sectors so that it becomes analytically difficult to follow through the impact of these elements.

Although many of the elements covered by computer-aided manufacturing are flexible, this flexibility nevertheless has its limits. It is often only through the interlinkage of these systems that a much wider-ranging flexibility can be attained. The key element in providing the necessary interlinkage of various flexible machines is the industrial robot (e.g. in automatic warehousing, inspection systems, machining centres) which also has its own built-in flexibility. It is around this nucleus that separate pieces of technology can be concentrated and integrated. Thus

"labour productivity in the engineering industry depends on automatic inter-operational transport and handling between the various production machines (for NC-machines). Industrial robots and manipulators seem to meet these requirements because they can handle a large number of operations which have so far been carried out by man within these integrated manufacturing systems."(6)

The industrial robot has been described as "... a

---

6. Economic Commission for Europe, Working Party on Automation, Seminar on Industrial Robots and Programmable Logical Controllers, Automat/Sem. 5/5 December 1977, page 2.

necessary but incomplete instrument in computer assisted manufacturing, and forms the centre point for development, on which a good part of the future of automated production is based, and around which a new production method can be generated"(7). The industrial robot is also perhaps the most visible element of CAM in that it is often a direct replacement for human labour.

In examining industrial robots and their applications it is necessary to recognise the interlinkages existing between the diffusion of robots and other automated capital equipment which come under the heading of CAM(8). A major advantage in an analysis of industrial robots is the relative ease in defining more exactly the industries and workposts most impacted by robots. As regards the future generation of industrial robots, the differentiation between these machines and other CAM elements is likely to become more distinct when the new generation of assembly robots is developed. When considering robots, the similarities and differences between these machines and other automated manufacturing equipment, in terms of machine characteristics and their applications, must be maintained in perspective. For example, as has already been noted in continuous process production, industrial robots have less relevance than computer integrated manufacturing systems. In production of discrete parts and low volume batch production robots have greater relevance(9).

Definition of industrial robots

Within the range of automated manufacturing equipment robots and the science of robotics refers to a specific area. The increasing flexibility, sophistication and application of automated processes has led to much of the literature using the term "robotics" loosely, often as a synonym for the range of automated manufacturing equipment. It is therefore useful to have a more restricted and exact definition of industrial robots, which is

---

7. Le Quément, J., The Social and Economic Stakes in International Competition in the Robotics Industry, 12th International Symposium on Industrial Robots, 9th-11th June, 1982, Paris; Hermès Publishing.
8. At present investment in industrial robots is a small part of total expenditure on automation, both "hard and soft". But the impact of robots is expected to be greater than their relative ranking in terms of expenditure because of the potentially greater flexibility and use in assembly as well as wider role and "machine-integrating" role.
9. According to J.F. Engelberger "the essence of robotics is to provide a form of automation which is immune to obsolescence. A proper industrial robot is neither product, nor industry limited." Robotics in Practice; (New York: AMACOM, 1980), page 17.

relevant to the application of robots to manufacturing industries(10). It should be kept in mind that robots have evolved from developments in machine tools, manual manipulators and remote manipulators. Their development has been evolutionary and has followed developments in the capital goods industry in general.

Robot devices have been defined by the Department of Industry in the United Kingdom as "reprogrammable mechanical manipulators". This definition is similar to that of the Robot Institute of America, that is:

> "A robot is a reprogrammable, multifunctional manipulator designed to move material, parts, tools or specialised devices through variable programmed motions for the performance of a variety of tasks."

This definition, which has become widely accepted, is used in the present study(11).

A robot consists of two basic parts: the mechanical structure (including the moving system) and the electronic command structure. The latter system provides the robot with its programmable memory as well as allowing for synchronisation with other industrial machinery. The ability to reprogramme adds to flexibility by allowing for a multiplicity of tasks, as well as in the ability to handle a variety of products, whereas other special purpose automated machinery is usually restricted to a single task or

---

10. The range in complexity of robots, the rapid evolution in the technology, and the merging of technologies which come under the heading CAD/CAM implies that over the medium term the various components including robots will be changing in definitional characteristics. The widening and merging of technologies into automated production systems is one of the more important aspects of computer-integrated manufacturing.
11. The national co-ordinating Committee of the International Symposium on Industrial Robots has similarly accepted the definition of the Robot Institute of America. The International Standards Organisation has adopted a very similar definition:
> "The industrial robot is an automatic position-controlled reprogrammable multifunctional manipulator having several axes capable of handling materials, parts, tools or specialized devices through variable programmed operations for the performance of a variety of tasks."

The Economic Commission for Europe's Working Party on Engineering Industries and Automation has also agreed to accept this definition.

product. There are three types of drive systems used for robots(12):

     i) Pneumatic using compressed air - this drive is lightweight, fast and relatively inexpensive.
    ii) Hydraulic using compressed fluids - this drive is more expensive than pneumatic, provides more lift and is appropriate for use in hazardous areas.
   iii) Electric - this drive has the greatest lifting capacity, consumes the least energy and is the most expensive, both to buy and maintain.

The arm geometry of the present robot generation is important, determining, in conjunction with the wrist and hand (end effector) subassemblies, the sphere of influence a robot has in its ability to reach work pieces, use tools, lift, grip, etc. For the existing generation of robots the mechanical structure remains a significant part of the robot in cost terms, but is expected to decline relative to control components in the future generations of robots(13). The next generation will, to different degrees, emulate some human senses using artificial intelligence. In particular, visual and tactile sensing will be required to undertake the more complex assembly tasks.

It is also possible to classify robots according to various functional categories. Thus, following the Japan Industrial Robot Association's (JIRA) classification of industrial robots(14):

Industrial Robots

1. Manipulator - directly controlled by an operator.
2. Sequential Robot - manipulator which functions by following a pre-established sequence.
   i) fixed sequence - difficult to change sequence
   ii) variable sequence - sequence easily modified
3. Playback Robot - manipulator which retains in memory a sequence shown by operator.
4. Numerically Controlled Robot - manipulating robot which receives orders through numeric control.
5. Intelligent Robot - robot which is able to understand the functions required through sensor capacity and recognition.

---

12. Electrical drive is the dominant system used at present and is expected to maintain and perhaps increase its existing share.
13. A discussion of the cost breakdown is provided in Chapter III.
14. This classification of robots is by input information and teaching method.

Based on the American definition the "manipulator", and "fixed sequence" categories would not be classified as robots, but rather as automatic machines. The broader definition used by JIRA has tended to result in some non-comparability of statistics although these data can usually be adjusted to fit the American definition(15).

The manipulator and sequential robots can be considered as first generation robots which have limited flexibility as compared to second generation robots (playback and numerically controlled robots). The third generation or intelligent robot is differentiated from previous generations by its sensory ability and capability to react to changes in its surrounding work environment.

Robots can also be classified on a different, but still functional, basis:

Pick-and-Place Robot - applied to elementary operations for material handling (would not be generally accepted as a robot under the present definition).

Servo Robot:

- programmable robot: a servo robot directed by a programmable controller that memorises a sequence of arm and gripper movements;
- computerised robot: a servo robot controlled by a computer which can receive new instructions electronically;
- sensory robot: a computerised robot with one or more artificial senses;
- assembly robot: a computerised robot, usually sensory, designed for assembly-line jobs.

Further classifications of robots are possible by categorisation according to work function, for example: surface treatment, handling, joining, transfer, heat treatment, assembly.

_____

15. Gradually, the US industry is moving in the direction of a compromise involving the inclusion of both manipulators and fixed sequence robots in robot production data and the breakdown of this total to show manipulators, fixed sequence robots, and robots as defined by the Robot Institute of America.

Chapter II

PRODUCTION AND SUPPLY ASPECTS

1. INTRODUCTION

The robot-producing "industry" is too young and in-
sufficiently homogenous to be classified as a separate
industry. Many of the producers are in the mechanical and
electrical engineering industries(1), particularly machine
tool companies which have diversified into robotics.
Often these industries are themselves important users of
industrial robots. Other producing firms are also robot-
using industries (in particular the automobile producers)
which, having identified a need as well as ways to use
robotics, developed them to their own specifications.
There is also an increasing trend in the establishment of
independent producing companies.

There are substantial differences in the structure of
robot production in the various Member countries. Thus
producers in the United States have tended, for the most
part, to be independent and specialised while the major
European and Japanese producers have tended to be integrated with larger companies. However, as the industry
expands this firm structure is changing rapidly, particularly in the United States, but also in Europe and Japan
where many of the medium and smaller enterprises specialising in robot production are increasing their market
share.

From the national strategy point of view there are
two main principles which have led governments to encourage, in various degrees, the domestic production of industrial robots and other computer-aided manufacturing
elements. Firstly, there is the belief that domestic production capability of new technologies provide a country
with the ability to master the evolution, diffusion and

_____

1. In many countries, in fact, the distinction between the electrical and mechanical engineering sectors is
becoming less apparent as these sectors merge into what
the Japanese refer to as the "mechatronics industry".

effective application of these technologies. Thus the benefits which such technologies bring can be maximised and in particular the ability to increase competitive-ness. Such a strategy in turn implies the need to have a role in research and development in these new technolo-gies, particularly to ensure that development is relevant to the domestic industrial structure. These strategic considerations are not restricted to national governments, but are viewed as relevant to firms which are likely to be major users of industrial robots. In particular, the need to control technological development is viewed as impor-tant by major users.

Secondly, the direct employment impact resulting from the introduction of industrial robots and other automated equipment is likely to be negative. Such a negative impact is usually immediate, occurring as equipment is introduced, but may be reduced somewhat, in the longer term, through the production of industrial robots and em-ployment creation arising from R&D activities, marketing, etc. which are needed to support production. Dependence on imports is generally viewed as implying an outward em-ployment leakage, as well as outward capital flow and technological dependence. A further, but important, aspect is the role of local suppliers in stimulating domestic demand. In order to stimulate growth, domestic suppliers need to make the capabilities of robots known to potential users, thus tending to stimulate applications.

2. COUNTRY PROFILES

The general situation discussed in the previous para-graphs can best be illustrated by a brief consideration of the structural evolution of the producing industry in some countries and also by reviewing some firm and national strategies concerning industrial robots(2).

*
* *

United States

Production of industrial robots in the United States has generally been based on independent, non-user, produc-ing companies which have focused on marketing and

---

2. National policies regarding robots are examined in Chapter IV.

development for customer requirements. The two largest producers of robots are Unimation(3) and Cincinnati Milacron (CM). Unimation had a market share in the United States of about 44 per cent in 1980-81, which declined to about 32 per cent in 1982; in Europe, where it has also diversified, Unimation has about 25 per cent of the market, while in Japan, Kawasaki Heavy Industries produce Unimation robots under licence. Unimation robots will also be produced in Finland for the Scandinavian and East European markets. Unimation's share of the world robot market is estimated at around 15 per cent. CM's share of the US market was approximately 32 per cent for 1980-81, declining to 18 per cent in 1982. The turnover of Unimation is derived 100 per cent from sales of industrial robots, while about 5 per cent of CM's turnover is attributed to robot sales.

The structure of production in the United States is rapidly changing. Many companies which are expecting to become large robot users are establishing themselves as producers for their own use as well as for general sale. These companies, notably General Motors, IBM, General Electric, Bendix and Westinghouse, should have a significant impact on the structure of production and the strategy of existing firms(4). A case in point is the recent acquisition by Westinghouse of a majority interest in Unimation and perhaps the eventual merger of the latter company as a wholly-owned subsidiary of Westinghouse. Many of the larger companies are entering into production by establishing divisions within their companies to develop expertise in robots and other forms of automation, and by entering into agreements with foreign (usually Japanese) producers. These agreements include technology exchanges, exclusive and non-exclusive marketing rights, often followed by licensing for manufacturing facilities in the United States. Many companies have also entered into agreements with several overseas producing companies providing them with a wide range of robots for sale and/or manufacture suitable for a variety of different industrial activities.

Examples of trends described in the previous paragraph include: (i) GM and Fanuc have established a joint

---

3. The world's first robot producer, Unimation, was an independent producer and has been the leading robot company for many years; its first commercial robot was produced in 1961. It has been estimated that in 1980 about 20 per cent of the world's robot population were "Unimates".
4. It is significant that a company such as GM is becoming a producer of robots with the intention of making general sales. While GM has manufactured capital goods in the past, these were mainly for the company's own use. The increasing importance of software is a prime motivator for the entry of IBM into the robot market.

venture under which Fanuc will share its robot design technology with GM and the two companies will jointly establish a plant in the United States; (ii) General Electric will produce and market a robot developed by an Italian firm (DEA). GE has also agreed to exchange robot technology with Volkswagenwerke AG enabling GE to market and manufacture VW designed robots. GE has also a technology exchange, marketing and production agreement with Hitachi; (iii) Westinghouse Electric Corporation has signed agreements with Mitsubishi Electric Corporation and Komatsu of Japan, and with Olivetti of Italy to market their robots in the United States; (iv) IBM will market Sankyo Seiki robots in the United States until it produces its own robots domestically; (v) Yaskawa Electric Manufacturing and the Bendix Corporation have entered into an agreement for production and sales tie-ups. Bendix has also acquired a share of Comau of Italy. All of these agreements between United States, Japanese, and European producers and/or robot users are illustrative also of the fairly close co-operation in technology transfer among robot-producing companies.

The expected high growth in demand for industrial robots in the United States during the 1980s (from about $200 million at present to $2 billion in 1990) and the need to acquire technology and production experience for the future, have been obvious contributing factors to the fast evolution in the American producing industry. For many of the potentially larger using companies (GM, GE, Westinghouse, IBM) the need to have easy access to equipment which closely meets their requirements is an important factor. Another factor of interest is that many companies are supplying a fairly wide range of robots based on a strategy that a user prefers to buy from the same supplier to ensure compatibility and better servicing. While there has been a tendency in the United States for some major users and potential producing companies to purchase smaller companies specialised in automation or electronic engineering (General Electric is a particular example), this has not yet become a significant trend.

The structure of production in the United States is also being influenced by the considerable number of foreign industrial robot producers who are establishing footholds in the US market either independently or through established American companies. Examples include the following: (i) The major Swedish producer, ASEA, is establishing a production facility in the United States; (ii) KUKA of West Germany has acquired manufacturing capability in the States; (iii) ACMA-Renault has set up a joint company with Ransburg of the United States for domestic production.

While it is difficult to predict the evolution of the producing industry in the United States, it seems clear that the present market leaders will lose their relative

dominance in terms of market shares as the industry grows
and becomes more diversified. The large number of new
entrants into the industry from various sectors will tend
to stimulate demand both because many of these companies
are themselves large users but also as producers start
pushing their products in the market. Given the diverse
interests of many of the new entrants, the market base is
likely to broaden to include a wider sectoral coverage,
and a wider range of applications.

Japan

Japan's production of industrial robots has increased
rapidly since the early 1970s so that it is now the fore-
most producer and user of robots, as well as the world's
leading producer of computerised automation in general.
From a production of 269 robots (narrowly-defined) in 1976
output has grown to 3,150 robots in 1980 averaging 85 per
cent growth per year in volume [Table II.1]. Initially,
Japanese production was concentrated on less sophisticated
first generation robots. However, substantial developments in technology have taken place by producing companies so that the share of more sophisticated robots in
total production has risen over the period 1976 to 1980
from 4 per cent to 16 per cent in volume terms. As a
result Japanese companies are now in a position to provide
technology to foreign countries.

The large internal Japanese market for industrial robots, in comparison to other industrialised countries, and
the growth in this market during the late 1970s, played an
important role in stimulating domestic production and thus
in Japan's leading role as a robot producer. One
consequence of this rapid growth is that a significant
percentage of the world's stock of industrial robots is in
use in Japan. Japan's robot industry is characterised by
a wide dispersion of production and many firms. At present, about 100 firms are involved in the production of
robots(5), about half of these medium and small-sized
enterprises. Many of the enterprises involved in robot
production are thus larger companies, mainly in the engineering industries, whose robot divisions are integrated
into much larger manufacturing groups which are themselves
frequent users of robots. Some of the new entrants are
also companies which are not in the engineering or electrical sectors, but have entered the industry beause they
view it as a high growth sector and therefore as a potential profit centre in the future. Given the number of
firms, and in spite of the fast growing market, domestic
competition is fairly intense among producers in order to
maintain and/or increase their market shares. At present,

_____

5. Based on a narrow definition of robots about 80
firms are involved in production.

Table II.1

JAPAN: INDUSTRIAL ROBOT PRODUCTION, BY TYPE(1)

Table II.1(a)

"Sophisticated" Robots	1976 Units	1976 Ymn	1977 Units	1977 Ymn	1978 Units	1978 Ymn	1979 Units	1979 Ymn	1980 Units	1980 Ymn
Playback Robot	183	1,999	357	3,761	506	4,373	662	6,653	2,027	15,785
N.C. Robot	6	52	11	85	25	273	89	1,745	992	21,795
Intelligent Robot	80	850	199	2,361	255	2,817	788	3,698	131	2,160
Total(2)	269	2,901	567	6,207	786	7,463	1,539	12,096	3,150	39,740
Annual Average Increase %	-	-	111%	114%	39%	20%	96%	62%	105%	228%
Total (based on a broad definition of robots)(3)	7,165	12,268	8,613	18,246	10,100	25,399	14,535	40,398	19,873	75,361
Share of sophisticated robots as % of total production broadly defined	4%	24%	7%	34%	8%	29%	11%	30%	16%	53%

Source: JIRA.

Table II.1(b)

Percent Distribution of Units

	1976	1977	1978	1979	1980
Playback Robots	68.0	63.0	64.4	43.0	64.3
N.C. Robots	2.2	1.9	3.2	5.8	31.5
Intelligent Robot	29.8	35.1	32.4	51.2	4.2
Total	100.0	100.0	100.0	100.0	100.0

Source: JIRA

1. Japanese data may vary since they are often quoted on a fiscal year basis; the above data are for calendar years.
2. Variable sequence robots are not included in this total.
3. The broad definition includes manual manipulators and fixed sequence robots.

because of the small market relative to its potential size, the intense competition to gain market shares, and some "learning curve" pricing, robots add only a small proportion to total turnover of the larger integrated enterprises and do not yet, in many cases, represent a profit centre.

Present strategy by some of the large producing firms in Japan is to develop and utilise robots within their own plants in order to increase their own productivity. Examples of firms following such a strategy include Hitachi - one of Japan's largest general electric machinery producers - which is embarking on a R&D programme for an "intelligent robot" to undertake sophisticated assembly tasks within its own plants(6).

Similarly, TODA Construction of Japan has launched a project in co-operation with a leading robot producer (Kawasaki Heavy Industries) to develop an "intelligent robot" for use in construction-engineering jobs. The project is aimed in particular for use of robots in nuclear power plants (building and dismantling) as well as in other construction jobs. Fanuc, a major numerical control tool manufacturer, is developing and using robots for use in its plants to produce parts of numerically controlled machine tools as well as robot parts.

Producers have primarily focused on supplying the fast growing internal market so that exports have represented only a small percentage of total robot production (approximately 2 per cent). However, export considerations for the future have been taken into account in the many technical co-operation and commercial agreements which exist between Japanese companies and their Western European and United States counterparts. The number of such agreements taking place between Japanese and European and US companies is in fact surprisingly high and perhaps indicative of the importance that companies are placing in the future impact of industrial robots and the need therefore for technological experience and expertise in the area. Although many of the agreements entered into between companies (US-Japan, Europe-Japan) call for domestic manufacturing facilities, these will initially lead to a fairly rapid growth in exports since many agreements call for the marketing during the first few years of the licensors' products on an original equipment manufacturing basis(7).

---

6. See also Chapter IV page 63.
7. Because many of the agreements between major Japanese robot producers and foreign companies envisage joint manufacturing facilities in various countries, it is not clear how this will affect the magnitude of export growth. The industry (through JIRA) forecasts total

(Continued on next page.)

In terms of size, Kawasaki is the largest builder of industrial robots in Japan. In 1981 it increased output by 30 per cent over 1980 to 600 units. The next largest builder is Yaskawa Electric Manufacturing Company at around 600 units, followed by Hitachi, Mitsubishi, Kobe Steel and Fanuc. Foreign penetration of the Japanese market has been mainly by Unimation of the United States through its licensing agreement with Kawasaki. ASEA of Sweden has plans to directly set up production facilities in Japan.

Western Europe

In Europe, as in Japan, many large-using companies are themselves large producers of industrial robots. The leading producing countries are Sweden (the third largest producing country after Japan and the US) and Germany. Other important producing countries are Norway, the United Kingdom, France and Italy.

The European Market is shared as follows:

		1979	1981
Unimation	(United States)	23%	23%
Trallfa	(Norway)	17%	20%
ASEA	(Sweden)	16%	15%
Electrolux	(Sweden)	10%	5%
Kaufeldt (Arbot)	(Sweden)	9%	4%
Volkswagen	(Germany)	8%	12%
Others		17%	21%

Source: Frost and Sullivan, Veckans Affärer and unpublished material.

Sweden

Sweden is the leading European robot producer with a share of about 24 per cent of the Western European market in 1981, as well as being the only European country with a large international presence in robot sales. It should be noted, however, that the market share in 1981 already represents a significant decline since 1979 when this share was 35 per cent. Sweden's leading position can be largely attributed to the strategy of the major domestic producer, ASEA, and aided by the high internal rate of

_____

(Note 7 continued from previous page)
exports of 46 billion yen by the mid-1980s and 88 billion yen by the end of the decade, that is 16 per cent and 17 per cent of anticipated total output. On the other hand, government forecasts call for a much lower rate of growth.

diffusion of robots. Other Swedish producers include
Electrolux, Kaufeldt and Atlas Copco. The ASEA group
recently acquired ownership of the robot interests of
Electrolux giving the company a share of one-fifth of the
European market. ASEA has also about 7 per cent of the
United States market with sales there of $9 million in
1981 and has begun to assemble robots in Spain mainly
aimed at obtaining a share of robot demand by the five
automobile producers established there; ASEA will also
begin in 1983 to assemble robots in France. The company
is actively trying to establish production facilities in
Japan. ASEA is presently a market leader in sales of arc
welding robots and is among the top five producers for
spot welding robots.

The Swedish market is as follows:

	1976	1977	1978	1979
	(units)			
Domestic Production	150	275	315	460
Exports	65	120	190	245
Imports	15	25	40	54
Apparent Consumption	100	180	165	269
Export Share (units) %	43	44	60	53

Source: Report of the Computers and Electronics Commission. Data for 1979 are not actual.

Robot exports form a large, and increasing, share of
domestic production (53% in 1979 in terms of number of
units and 60% in value terms).

Germany

Germany has about 20 domestic producers, the largest
of which are Volkswagen followed by KUKA. Large electrical engineering companies such as Siemens have set up subsidiary companies for production of robots. Sales by
Volkswagen are mainly internal to the company so its large
market share, about 20 per cent, reflects its own usage of
robots (about 800 at present increasing to 2,000 by
1990). Other market leaders are KUKA followed by
Unimation (US) and Trallfa (Norway). Total sales of
robots in 1980 in Germany and their distribution between
imports and domestic manufacturing were as follows:

	Sales (units)	Imports	Own Manufacture
1980	1,133	49.5%	50.5%

Source: British Robot Association.

France

In France there are about 12 producer companies. The companies tend to be small and specialised. Renault is the exception which, through its subsidiary ACMA, is a major producer. Renault takes about 50 per cent of its own production and the remainder of deliveries are mainly to other automobile producers including Peugeot, Citroen in Spain and Volvo in Belgium. ACMA's joint venture in the United States indicates that the company intends to expand to try to obtain a share of the important United States market. Few joint ventures between French and foreign companies seem to exist(8). The French robot market was about 100-150 million francs in 1980, growing to approximately 300 million francs in 1982. Import penetration is high, estimates indicate that 50 per cent of the existing robot stock is imported. The government has recently given increased priority in its aim to develop a strong manufacturing base for robots.

Italy

Italy has four major producers and many smaller companies, some of which are highly specialised. The production share is divided between Basfer (16 per cent), Olivetti-Osai (7 per cent), Norda (55 per cent) producing mainly manipulators and Comau-Fiat (8 per cent). Sales by Fiat and Olivetti initially were mainly internal to the firms but are now increasingly being placed on the market. Domestic production expanded rapidly between 1975-1979, averaging 63 per cent per year and significantly higher in value terms (see Table II.2 below). Exports of robots also grew rapidly, accounting for nearly 40 per cent of the value of production in 1979. Imports declined from 78 per cent of apparent consumption in 1975 to 16 per cent in 1979.

Norway

The Norwegian company, Trallfa, has a substantial market share for painting robots in Europe, in Japan and in the United States. Presently about 80 per cent of the world's stock of painting robots are estimated to be Trallfa robots. In the United States Trallfa has a manufacturing and distribution agreement with DeVilbiss and in Japan with Kobe Steel. The company is a good example of a small company, and small country, specialising in a

---

8. Recently an agreement was signed between Yaskawa of Japan and the Compagnie Electro-Mecanique (CEM) to market robots of each company in Japan and France as well as for Yaskawa to provide technical assistance to the French company.

Table II.2

ITALY: INDUSTRIAL ROBOT MARKET

	1975	1976	1977	1978	1979	Average Annual Growth
Quantity (units)						
Production	60	120	160	270	420	62.6%
Imports	40	60	120	40	40	-
Exports	20	30	60	90	160	68.1%
Apparent Consumption	80	150	220	220	300	39.2%
Value (lira billions)						
Production	0.7	4.0	5.7	11.3	20.0	131.2%
Imports	1.8	2.8	6.4	2.2	2.3	6.3%
Exports	0.2	0.4	1.5	3.1	7.8	151.6%
Apparent Consumption	2.3	6.4	10.6	10.1	14.5	39.4%

Source: Le Progrès Scientifique.
[These data based on a broad definition include manipulators. Approximately 50 per cent of 1979 production is estimated as robots narrowly defined.]

product and becoming a leader in its field.   The only
other Norwegian robot-producer which has a smaller production capacity of point-to-point robots is Jonas Oeglaend.

One reason for the relatively high market shares of
Trallfa is the fast growth in the use of industrial robots
in the automobile industry for surface treatment. Although potential demand in the automobile industry for
industrial robots remains high, other areas of application
are growing in importance so that relative market shares
may be expected to change rapidly in the future.   In anti-
cipation of this shift in applications Trallfa has recently embarked on a diversification strategy.

United Kingdom

The United Kingdom has a small production base in
industrial robots with only about 8 producers. As a
result, imports form a large proportion of the robot stock:

	1980	1981	1982
Domestically produced	23%	30%	25%
Imports,	77	70	75
of which: Japan	4	8	14
USA	39	30	24
Europe	34	36	37

Source: British Robot Association.

27

However, the ratio of domestic production to sales is expected to improve over the next few years(9). This will result from plants which have recently come into production, such as that of Unimation, and new production facilities which are to be established from various new joint agreements (e.g. Dainichi Kiko and the Sykes Group, the 600 Group and Fanuc, and GEC Electrical Projects and Hitachi). Many smaller firms are also rapidly entering into production. The domestic market in 1981 was about £20 million but declined to £15 million in 1982. Government strategy has emphasized applications as the major priority.

Other Member Countries(10)

The manufacturing base for industrial robots in the other Member countries is either small or non-existent. In countries where there is some production of robots this is mainly assembly through licensing arrangements with one of the larger producers.

3. CONCLUSION

The high rate of entry of new firms into the robot industry is significant, notably by larger companies which are expected to be important users as well as having the financial strength to carry out longer-term research and development programmes. Many of these firms are engaged in areas where industrial robots are expected to have the greatest impact on manufacturing. The entry of major multinational companies is also indicative of the important role being attributed to robot technology over the next decade.

Despite the anticipated high rates of growth in demand for robots in the 1980s, it can be expected that as the industry matures there will be some mergers,

---

9. In 1981, of the net additions to the stock, 30 per cent of robots were of British origin.
10. Note should be made of efforts of the COMECON countries to introduce industrial robots and to obtain the relevant technology from OECD countries, notably the US, Germany, Italy and Japan. In some COMECON countries the low degree of labour productivity is possibly the most important aspect stimulating demand, but in countries such as East Germany and Czechoslovakia labour shortages are the prime factor stimulating applications. The Soviet Union's significant efforts to increase the utilisation of robots is primarily motivated by low labour productivity, absenteeism and skill shortages.

consolidation and the emergence of multinational producers, as well as some degree of specialisation. The industry structure is likely to resemble that of the machine tool industry with some large companies as well as many smaller specialist producers. Economies of scale are unlikely to be important in the industry in terms of production, but firm size may be important as concerns financial strength, marketing outlets and capability to undertake longer term research and development. In particular, the ability by firms to provide comprehensive automated production "packages", in terms of robots integrated with other automated manufacturing equipment including CAD/CAM and software services, is expected to become more important in the future. In the shorter term, competition for market shares may lead to substantial reductions in the price of robots.

There appears to be a greater dynamism by Japanese companies compared to companies in other Member countries, in realising the importance of robot technology. As a result, potential users are quicker in applying this new technology, thus stimulating production. Of particular importance are the efforts being made by Japanese producer-user companies in ascertaining their own future needs for robots, especially in assembly and undertaking intensive R&D programmes in order to be in a position to introduce flexible automation for assembly by the mid-1980s. The increased entry of large American firms into robot production over the last few years, many of them large potential using firms, can be expected to significantly stimulate production and diffusion over the next decade. The situation in many European and in other OECD countries contrasts sharply with trends in the United States and Japan. This is particularly the case in some European countries where diffusion and the entry of new producing firms have been much slower.

One aspect which may be of concern in some countries is that the dominant firm which emerges is also a dominant user (e.g. Renault in France, VW in Germany, Fiat in Italy). Such a production structure can retard the spread of technology, and slow down diffusion by not encouraging applications of potential non-user firms. For example, in the case of ACMA (Renault), Comau (Fiat) and VW, although they are major producers of spot welding robots their impact on the "open" market has not been significant.

Chapter III

TRENDS IN APPLICATIONS AND DEMAND

1. INTRODUCTION

The growth of the robot industry is mainly determined by the relationship between the cost of robots, the use they can be put to, and the expected payback period. While many analysts believe that present prices are high and are a deterrent to sales, the increasing sophistication of the technology increases the flexibility of the machinery and therefore alters the investment evaluation. One example of increased flexibility is the reprogrammability of robots which reduces their obsolescence.

At present, the economic recession and high cost of capital has led to a general reduction or postponement in investment plans. The effect of this macroeconomic situation on investment in robots, and other automated systems in general, is ambiguous. It is usually the case in a slow growth situation that the investment priority of firms is in reducing costs, improving factor productivity and rationalising production; thus investment in robots and flexible automation would be attractive. At the same time, however, the shortage of investment funds may place a constraint on many potential investors, in particular small and medium enterprises. In this case, it can be expected that, as the economic climate improves, the introduction of robots in manufacturing will accelerate. Probably the most important factor in stimulating the uptake of robots will be the increased awareness by firms, especially those striving to remain at the same technological level as their competitors, of the potential robots offer in terms of improved productivity. Awareness of this potential is apparently accelerating sales before an economic upswing becomes evident. Demand is likely to become more sensitive to macroeconomic forces in later years when market penetration rates are significantly higher.

While the demand for and the application of robots are interrelated, it is nonetheless useful to distinguish between the two for analytical purposes. Both applications and trends will depend on the success of the new

generation of robots particularly in the development of
visual and tactile sensory capabilities which will allow
robots to be introduced into many assembly tasks.

II. APPLICATIONS

Present manufacturing techniques emphasizing mass
production and specialisation have led to work procedures
which are simplified and repetitive, facilitating the introduction of robots which can easily be programmed to
perform such tasks. The present trend in the application
of robot installations has been mostly in work positions
which are hazardous, pose health risks or are monoto-
nous(1). Examples include spot welding, spray painting,
stamping presses, forging applications and palletizing
applications. The scope of applications is quickly widening as newer, more intelligent robots with greater flexibility are applied to a range of positions including assembly tasks, and as adaptation problems to a range of
workposts are overcome.

Main areas of application of industrial robots are
summarised below(2).

    i) Die-casting - this industry has been one of the
       first to use industrial robots. Two factors of
       importance facilitated the introduction of indus-
       trial robots to this industry. Firstly, "con-
       sistent and precise part orientation is essential
       in using robots, and this is inherent in die
       casting"(3). Secondly, the ancilliary equipment
       required is minimal when introducing robots to
       die-casting operations. The unpleasant working
       environment associated with die-casting has faci-
       litated applications.

   ii) spot welding - the first spot welding line was
       introduced in 1969 for automobile production and
       is the main area of application of the present
       population of industrial robots. Most of the
       spot welding robots in the automobile industry
       work on a 3-shift basis and are capable of

---

1. A wide range of robots, broadly defined, are being
used and developed for use in non-manufacturing applications including nuclear power stations, undersea explora-
tion and hydrocarbon production, medical applications,
constructions, agriculture, mining, etc.
2. This paragraph draws heavily on J.P. Engelberger:
Robotics in Practice, op. cit.
3. ibid page 145.

adjusting to various body styles(4). As a rule of thumb, one spot welding robot displaces two manual workers per shift.

iii) arc welding(5) - this application requires little change in existing equipment; the robot uses the same welding gun used by a manual worker. Automated arc welders are in use, but they tend to be less flexible, while a robot arc welder allows it to be used where variable products are manufactured.

iv) investment casting - the increased requirement for intricate and accurate speciality parts has resulted in the considerable growth of the investment casting industry. Industrial robots are used in casting moulds and in the transfer of parts between work stations.

v) forging applications - despite the hazardous work conditions in forging shops, automation has been difficult because of batch production runs. Industrial robots have been applied in this industry but only to marginal functions since they cannot as yet handle the complexity of some operations (particularly where speed or visual judgements are required). Applications are mainly where the requirement is to handle hot metal parts, to position those parts and to respond to signals from other equipment.

vi) press work applications - (automobile body panels and automobile parts, aircraft structures and many of the domestic appliance industries use press operations). Many press operations, particularly where large parts are involved, are partially automated either through standardized "hard" automation or through the use of limited sequence machines (included as robots in the wider Japanese definition). For smaller parts with a shorter handling cycle the disadvantage of robots is their relatively slow speed. Robots have made a marginal impact in press work applications. Many major industrial robot producers (Unimation, Cincinnati, Volkswagen) are presently working on extending applications in this area. Safety considerations favour the introduction of industrial robots in press work applications.

vii) spray painting and surface treatment - hazardous work conditions have been a prime reason for

---

4. While special purpose automated machinery can be used for spot welding, such machinery is only economical when no product design changes are envised.
5. In arc welding it has been estimated that robot productivity is three times greater than that of manual welding with a required 4 to 5 years for amortization of the investment.

introducing automation to spray painting applications. Although automated spray painting
machines are in use their flexibility is limited and they are unable to cover less access-
ible areas (which in the automobile industry is
an important requirement) so that the use of
industrial robots with their flexibility in
spray painting is expected to grow rapidly,
particularly in the automobile industry. The
changes in car style and colour can be accommo-
dated easily by the robot's flexibility. Other
areas which use robots for surface treatment
include the white goods industry, furniture and
cabinet industries which use robots for var-
nishing and painting, and areas requiring surface cleaning, porcelain coating, etc.

viii) plastic moulding - robots are in use in unloading injection moulding machines, trimming
mouldings, loading moulds, and palletizing and
packaging mouldings. The economic justifica-
tion for installations has been based on labour
saving, higher productivity and fast cycle
times. Poor working conditions have also been
a factor stimulating installations.

  ix) foundry applications - foundry operations are
characterised by hazardous work conditions, a
shortage of skilled workers, and the requirement for visual adjustments and judgements by
an operator. Although automation is desirable
in foundry work, it is not always easy, and as
a result the penetration of industrial robots
has been fairly slow. Workposts where robots
have been applied include ladling hot metal
into moulds (at present a marginal area of
application), mould care, unloading machines,
and in some fettling operations.

   x) machine-tool loading - robots are increasingly
playing a central role in machine-shops. As
Engelberger notes "the objectives of automation
in the machine shop, as elsewhere, have always
been to reduce in-process inventory, cut lead
time, minimise direct and indirect labour
costs, make maximum use of capital equipment
and generally increase the number of acceptable
parts made per shift. Often these aims had to
be pursued in the face of regular product
change which made general purpose automation
difficult to apply"(6). The increased sophistication in control technologies for advanced
machine tools (lathing, drilling, milling and
grinding machines) has increased their accuracy
and flexibility in the machining of various
parts. While "hard" automation can and has

6. Engelberger, op. cit., page 234.

33

often been used to connect the various machines in a transfer line, an industrial robot transfer line can increase flexibility and reduce machine downtime. This is the case for small batch manufacturing as well as for the larger machining centres.

   xi) <u>palletizing applications</u> - in the manufacturing and distributive trades industrial robots have been found particularly useful in palletizing and depalletizing.

  xii) <u>assembly applications</u> - in assembly, robots are used for electric/electronic machinery and apparatus, automobile electrical equipment, mechanical engineering, toys, etc.

The areas of application mentioned above can also be classified according to the type of work undertaken, e.g. robots used in handling (machine or tool handling, plastic moulding), surface treatment (painting, cleaning), joining (spot welding, arc welding), heat treatment (forging, investment casting), transfer and assembly. The areas of application described above are not an exhaustive list but indicative of the type and variety of applications of industrial robots. They also indicate the flexibility of industrial robots in that they can be introduced into manufacturing systems (e.g. machining centres), in flow lines (e.g. automobile spot welding) and in isolated work stations. One aspect in the development of robots which should be noted, however, is the increasing tendency to produce specialised robots (e.g. for painting, arc welding, etc.). While these remain flexible by being able to handle variations in a particular task, this flexibility is restricted to that task.

*
* *

The following sections examine some applications, firstly according to workposts and then by industry. The final section examines the demand trends for industrial robots in some Member countries.

Applications by Workpost

The first use of industrial robots was in the automobile industry, which in most countries has remained as the primary user. The main applications in this industry are in spot welding, followed by surface treatment applications, particularly painting. The rapid development of robot technology has led and is leading to a wide range of applications relevant to most industrial sectors and affecting most workposts. The rapid growth of robot applications and wide diffusion across industrial sectors is

34

also leading to a rapid change in the distribution of robot population by workpost. Thus by the late 1980s it is expected that applications in workposts related to spot welding, painting, and to metal working, will have declined in relation to assembly work.

The distribution of robots in Japan by workpost is similar to that of other robot-using countries with the exception that assembly applications are already fairly significant (see Table III.1). This is partly due to the widespread use of fixed sequence robots and manipulators, although there is now an increasing emphasis in Japan on high-technology robots. However, at present robot usage is limited to only elementary assembly applications. Applications of robots in Japan to welding, painting and

Table III.1

JAPAN: APPLICATIONS OF INDUSTRIAL ROBOTS

(Percentage distribution based on total value)

	1979	1980
Assembly	13.3	32.5
Painting	2.0	2.0
Spot welding	10.2	10.4
Arc welding	4.5	7.6
Others	70.0	47.5

Source: JIRA.

(Data are based on a broad definition of industrial robots, thus the distribution to assembly applications is exaggerated by including manipulators, etc.)

Table III.2

JAPAN: ROBOT PRODUCTION

Robot type	Percentage increase in value 1979 to 1980	Percentage increase in number 1979 to 1980
Assembly	340	33
Arc welding	211	200
Painting	88	100
Press processing	60	6

Source: JIRA

(Data are based on a broad definition of industrial robots.)

metal pressing posts have been explained by the shortage
of skilled labour in these areas and the need to amelio-
rate poor working conditions. The emphasis on productivity growth is the prime motivation stimulating applications in assembly workposts. The significant changes
taking place in the pattern of demand for different robot
types, based on production trends, was already evident in
1980 in Japan [see Table III.2]. The data also reflect
the increasing sophistication of industrial robots, given
that the growth in value was much higher than the growth
in the number of robots.

In the United Kingdom the use of robots has been
mainly for spot welding, surface treatment and injection
moulding. The installation of robots in these three areas
as a percentage of total installation is as follows (see
also Table III.3):

	1980	1981	1982
Spot welding	16%	23%	22%
Surface treatment	18%	12%	11%
Injection moulding	15%	12%	14%

Spotwelding applications, which increased fourfold between
1980 and 1982, have shown the highest growth in installations mainly reflecting the robotization process in the
automobile industry. The relatively high number of applications in surface treatment also reflects the use of
spray painting robots in the automobile industry. Other
high growth areas of application include arc welding and
machine tool servicing, where installations increased
threefold over 1980-82, and in assembly posts where applications remain very few, but nevertheless increased their
share from 1 per cent to 3 per cent of total installations.

At present, many robot installations in the United
Kingdom have been in isolated work stations rather than
placed together in a continuous production line. There
are few firms with multiple robot installation, the important exception being in the automobile industry where multiple installations are used mainly for spot welding.

In contrast to the United Kingdom, the applications
in Germany have followed a different configuration (see
also Table III.3):

	1980	1981	1982
Spot welding	30%	33%	38%
Handling	15%	16%	15%
Surface treatment	14%	10%	11%

Table III.3

ROBOT APPLICATIONS IN THE UNITED KINGDOM AND WEST GERMANY IN 1980, 1981 AND 1982

APPLICATION	United Kingdom						West Germany					
	1980		1981		1982		1980		1981		1982	
	No.	%	No.	%	No.	%	No.	%	No.	%	No.	%
Surface Treatment	69	18	88	12	124	11	155	14	231	10	397	11
Spot Welding	59	16	167	23	249	22	339	30	771	33	1,331	38
Arc Welding	48	13	87	12	157	14	138	12	227	10	585	17
Grinding, Drilling, etc.	4	1	6	1	13	1	5	--	10	--	20	1
Assembly	5	1	15	2	32	3	44	4	101	4	122	3
Press tools	21	6	24	3	27	2	28	2	25	1	70	2
Die Casting	33	9	33	5	36	3	56	5	113	5	120	3
Injection Moulding	54	15	89	12	167	14	113	10	--	--	--	--
Machine Tools	39	10	56	8	111	10	34	3	29	1	193	6
Manipulation	8	3	10	1	12	1	--	--	--	--	--	--
Inspection, Test	3	1	7	1	15	1	--	--	--	--	--	--
Palletizing, Packaging	16	4	24	3	38	3	170	15	374	16	--	--
Handling, General	2	-	3	-	10	1	17	1	22	1	520	15
Research	--	--	10	1	38	3	--	--	--	--	58	2
Others (unspecified)	10	3	94	13	123	11	34	3	398	17	84	2
Total	371	100%	713	100%	1,152	100%	1,133	100%	2,301	100%	3,500	100%

Source: British Robot Association. (Data for machine tool applications are not given in the 1981 BRA survey for Germany. These data may have been categorised under "handling, general".)

Table III.4

ROBOT APPLICATIONS BY WORKPOST IN VARIOUS OECD COUNTRIES(1)

CANADA (1981)	%	DENMARK (1981)	No.	%
Spot welding	48	Machine Loading	23	44
Spray painting	22	Arc welding	6	12
Material handling	19	Surface Coating	10	19
Assembly	3	Handling	8	15
Others	8	Others	5	10
Total	100	Total	52	100

FRANCE (1981)	No.	%	ITALY (1982)	%
Material handling	340	43	Welding (arc and	
Spot welding	300	38	spot)	37
Painting	70	9	Casting, handling	27
Arc welding	80	10	Machine tool	
Total	790	100	loading/unloading	18
			Surface treatment	9
			Assembly	9
			Total	100

NETHERLANDS (1982)	%	SWEDEN (1981)	%
Arc Welding	36	Handling and machine	
Spray painting	28	servicing	50
Deburring	12	Surface treatment	15
Loading/unloading		Spot welding	13
machines	10	Arc welding	10
Spot welding	4	Other processing	7
Others	10	Other	5
Total	100	Total	100

UNITED STATES (1982)	No.	%
Machine tool		
loading/unloading	1,470	24
Spot welding	1,190	19
Die Casting	880	14
Painting	290	5
Arc Welding	270	4
Other handling	1,950	31
Investment casting	120	2
Assembly	50	-
Finishing	30	-
Total	6,250	100

Sources:

Canada - Robotics in the Canadian Manufacturing Industries,
National Research Council Canada
Denmark - The Industrial Robot, Volume 9, No. 1 March 1982
France - BIPE
Italy - Italian Society for Industrial Robots, SIRI
Netherlands - Stichting Toekomstbeeld der Techniek
Sweden - Whalbin, C., A. Wallström, and A. Weinz, Swedish
industrial robot market - comparison with UK, The Industrial
Robot Volume 9, No. 4 December 1982
USA - Robot Institute of America.

---

1. It should be noted that the description of a workpost is not comparable in all cases.

The higher rate of penetration of robots in the German
automobile industry compared to the United Kingdom, as
well as differences in the relative sizes of the industry,
are the main reasons for the higher percentage of applications in spot welding in Germany. The number of spot
welding robots in use in Germany in 1982 exceeded in fact
the total robot population of the United Kingdom for that
year.

In Italy data indicate a high concentration of robot
use in welding (37%) and in metal forming applications,
particularly in die and investment casting, and handling
(27%) [see Table III.4]. In France the dominance of en-
gineering/automobile applications has meant that material
handling and spot welding applications have been the most
important (Table III.4). The distribution of robots in
Sweden varies somewhat from that in other countries with
an emphasis on handling and machine servicing and a lower
distribution in spot welding applications. In the United
States applications have been mainly in machine tool load-
ing/unloading, spot welding and die casting
(Table III.4). For Canada, robots were concentrated in
spot welding and painting posts, while in the Netherlands
applications are mainly in arc welding and painting workposts.

Applications by Industry

Broadly, the distribution of production within the
manufacturing sector of the major OECD countries follows a
similar pattern (Table III.5). In particular, the manufacture of machinery, the manufacture of electrical
machinery, and the manufacture of transport equipment is
important, accounting for about one-third of the value of
total manufacturing output. Thus it is not surprising
that in general the industrial distribution of robots is
fairly similar among countries [Tables III.6 to III.8].

As has been noted, the automobile industry is the
major user of industrial robots in most Member countries
which at present have a relatively large robot population. This trend is primarily a result of technological
developments in robotics which were initially concentrated
in spot welding and painting applications, both areas
being of importance to the automobile industry. At the
same time, the need for automobile producers to increase
productivity, increase quality and reduce wage costs also
stimulated robot developments of relevance to automobile
production. The industry can be expected to remain a
major user of industrial robots in the future.

The potential breadth in the industrial application
of robots is evident from Table III.6 which provides
detailed data on robot distribution by industry for the
United Kingdom for 1980. The automobile industry in this
country accounts for one-third of the robot stock,

Table III.5

SECTORAL SHARE OF TOTAL MANUFACTURING PRODUCTION

(Percentages)

Manufacturing sector	Germany	USA	France	Japan	UK	Sweden	Canada
ISIC	(1981)	(1977)	(1981)	(1978)	(1978)	(1980)	(1978)
3810 Manufacture of fabricated metal	5.0	6.0	3.3	4.6	6.9	6.8	8.2
3820 Manufacture of machinery	11.0	9.6	10.0	10.5	9.7	10.5	5.1
3830 Manufacture of electrical machinery	9.2	6.8	6.0	8.7	6.7	6.3	6.4
3840 Transport equipment	11.6	12.3	11.5	9.6	9.4	11.0	12.4
of which:							
3843 Motor vehicles	10.0	7.0		7.0	5.9		
Total (3810 + 3820 + 3830 + 3840)	36.8	41.7	30.8	40.4	38.6	34.5	32.1

Source: OECD.
Data are based on current prices and in national currencies, except for Canada which are based on value added 1970 prices.

Table III.6

ROBOT DISTRIBUTION IN THE UNITED KINGDOM BY INDUSTRY:
1980
(Based on Standard Industrial Classification)

Industry	No. of Robots	%
Energy and Water Supply (inc. coal and nuclear)	6	1.6
Metal Manufacture and Mineral Extraction	1	0.3
Metal Goods Manufacture	64	17.2
Mechanical Engineering	37	10.0
Office and Data Processing	2	0.5
Electric and Electronic Engineering	20	5.4
Automotive and Automotive Parts	126	34.0
Other Transport (inc. Shipbuilding and Aerospace)	20	5.4
Instrument Engineering	4	1.1
Textiles	3	0.8
Timber and Wooden Furniture	2	0.5
Rubber and Plastic Processing	52	14.0
Other Manufacturing	34	9.2
Total	371	100.0

Source: British Robot Association (1981).

followed by the metal goods industry and plastic
processing. In Sweden, the relatively low level of
applications of robots in the transportation industry
(22 per cent of the robot stock) reflects the fact that
automobile production is, in comparison with other
countries, relatively small vis-à-vis the transportation
equipment sector as a whole, and the metal working sector. The latter sector accounts for 50 per cent of robot
applications (see Table III.7), a significantly higher
proportion than in other industrialised countries. In
Japan [Table III.8] the electrical machinery sector became
the major user of robots in 1980 purchasing mainly the
more sophisticated robots (playback, NC and intelligent
robots). The high rate of application of robots in 1980
by the electrical machinery sector (1247 units) and the
automobile sector (1211 units) is indicative of the
emphasis being placed by Japanese producers on increasing
productivity and maintaining their international competitiveness in these sectors(7).

---

7. The additions to the robot population by these two
industries in 1980 were greater than the stock of robots
in Sweden and Germany together for the same year.

Table III.7

USE OF ROBOTS BY INDUSTRY IN VARIOUS OECD COUNTRIES
(Percentage distribution based on number of units)

CANADA (1981)		GERMANY (1981)	
Automobiles	63%	Transportation Industry	46%
Plumbing Fixtures	9%	Electrical Engineering	14%
Electrical Engineering	6%	Mechanical Engineering	12%
Metalworking	6%	Metal-Working Industry	6%
Appliances	5%	Plastic and Materials	
		Industry	6%
		Other	16%
ITALY (1979)		NETHERLANDS (1982)	
Automobiles	28%	Metal Working Industry	64%
Household Appliances	8%	Mechanical Engineering	12%
Metal Industries	8%	Electrical Engineering	9%
Electrical Industry	6%	Transport Equipment	5%
Rubber Industry	1%	Construction materials	3%
Exports	49%	Rubber and plastics	1%
		Others	6%
SWEDEN (1979)			
Metal Working Industry	51%		
Mechanical Engineering			
Industry	15%		
Transportation Industry	22%		
Electrical Engineering	9%		

Sources: Canada - National Research Council Canada, op cit.
Germany - KUKA, quoted in Wirtschaftswoche Nr. 15,
8.4.83.
Italy - Le Progrès Scientifique - Mars-Avril 1981
Netherlands - Stichting Toekomstbeeld der Techniek
Sweden - Swedish Electronics Commission

Table III.8

USE OF ROBOTS BY INDUSTRY IN JAPAN
(Percentage distribution based on total sales)

	1976	1977	1978	1979	1980
Automobiles	30	34	39	38	30
Electrical Machinery	21	23	24	18	36
Plastic Moulding	13	10	10	11	10
Metal Products	6	3	7	8	5
Iron and Steel	6	6	3	4	1
Others	24	24	17	21	18

Source: JIRA (based on a broad definition of robots).

In Canada, the concentration of robots in the automobile industry (50 per cent of the robot stock) reflects the fact that the manufacturing industry has in general been relatively slow in investing in robot applications. Industrial applications in Germany of robots have been concentrated in the transportation sector. In Italy, up to 1979, the take-up of robots by the automobile industry was weak, while for the Netherlands there is a heavy concentration of robots in the metal working industries.

One aspect of note in the application of robots in various countries is the concentration of users. Thus in Sweden, for example, robots are concentrated in a few large companies with the seven largest robot users accounting for 45 per cent of the total stock. A similar pattern exists in other countries including those with only a small robot population. This concentration is explained by the fact that automobile companies have been the major users in most countries as well as because most large firms use production lines and therefore can install multiple applications, and in integrated plants can apply different types of robots.

3. DEMAND TRENDS

Various factors are at work stimulating the demand for robots. These factors, which are in some cases interrelated, include:

a) Productivity improvement

The need to increase or maintain competitiveness, and thus improve productivity, is an important factor in increasing the speed of diffusion of industrial robots. This is, for example, the case in industries where international competition is strong, such as automobiles, machine tools, and electrical engineering. Although strategies vary by company and country, a prime factor is often the need to emulate the leading firm in the particular industry. An important aspect in the productivity impact of robots is that they assist in producing products of a high and consistent quality(8).

b) Labour costs

In many OECD industrial sectors, the low rates of increase in labour productivity coupled with rising hourly

---

8. The successful application of robots may require product modification, which in itself may be important in increasing productivity.

costs of labour have been viewed as the main reason for the rise in total costs per unit output, deteriorating profitability and declining competitivity. For many companies the increase in direct wage costs coupled with non-wage labour costs (social contributions, employment protection legislation, redundancy payments, etc.) have played an important role in stimulating interest in flexible automation. In contrast to rising labour costs, robot prices are remaining fairly constant and in some cases declining, so that payback calculations are increasingly favouring investment in industrial robots. The increasing trend in the reduction of working hours, increased paid annual leave, etc. will also tend to favour investment in robots (conversely, investment in robots and other forms of flexible automation will help realise in the longer-term social programmes aimed at lowering working hours, etc.). The question of labour reliability in some countries, either because of a high rate of absenteeism, or a large number of wildcat strikes or perhaps because of a lack of consistency in the quality of output, has also helped increase the attractiveness of investment in robots.

c) Improved work conditions

The increased recognition for the need to improve work conditions and safety standards, by eliminating manual labour from workposts which are health-hazardous, dirty or dangerous, as well as the costs involved in workmen's compensation, are playing an important part in stimulating the demand in applications of robots(9). For example, in Japan the Small Enterprise Loan Corporation assists small and medium enterprises to replace labour in areas which pose safety or health problems. The attempts by industry to improve work conditions has also been stimulated by the difficulty in obtaining and/or keeping labour in workposts which are dirty and dangerous.

d) Labour shortages

Despite present high unemployment rates, there are difficulties in many countries in finding certain types of skilled labour (welders, machinists are often cited as examples). In some cases the problem of skilled labour shortages can be overcome by introducing robots. From a longer-term perspective there are concerns in some

---

9. A recent Delphi survey by the University of Michigan and the Society of Manufacturing Engineers in the United States [Industrial Robots: A Delphi Forecast of Markets and Technology] estimated that installed robots will decrease factory accident rates by about 11 per cent in 1985, up to 24 per cent by 1990 and nearly 41 per cent by 1995.

countries (e.g. Japan) that because of the age profile of
the population as well as the increase in the level of
education of new entrants into the labour force, there
will be a shortage of unskilled labour in the future.
This factor is viewed as important in stimulating robot
applications in the future. The difficulty in inducing
labour to work third shifts is also a factor favouring the
adoption of increased automation.

e) Flexibility in production

Manufacturing concepts are de-emphasizing standard-
isation and assembly line techniques for more production
flexibility which facilitates the cheap and quick adoption
of engineering changes, changes in consumer tastes and
requirements and the ability to adjust easily to fluctua-
tions in consumer demand, and the shorter life-cycle
characteristic of many new products. The requirement for
more flexibility in production is relevant to many industrial sectors using assembly line production methods as
well as to sectors where manufacturing is based on batch
production methods.

In contrast to the demand stimulating factors above,
other forces act as a brake on demand. These include:

a) Robot Costs.

For the present generation of industrial robots the
cost in many applications and of some robot types is declining. However, for some robot types and in particular
in more sophisticated applications, robot costs can be
expected to increase. For example, in Japan average per
unit values for some robot types have been as follows:

AVERAGE PER UNIT VALUE (Y000)

	1977	1978	1980
Playback Robot	10,923	8,642	7,787
N.C. Robot	8,666	10,920	21,970
Intelligent Robot	10,625	11,047	16,488

Source: JIRA

Thus while prices declined for playback robots (which ac-
counted for over 60 per cent of new application in Japan
in 1980), the high technology robots increased in price.
As robots become more sophisticated the relative changes
in cost distribution will also affect prices. An estimate
of the cost breakdown is given below:

PROPORTIONAL COST BREAKDOWN

	1980	1985
Mechanical Part	15%	10%
Electronic Part	15	15
Drive	20	15
Visual and Pattern Recognition	5	15
Sensors	15	25
Minicomputer control	30	20

Source: Le Progrès Scientifique, Mars-Avril 1981.

In contrast to the actual cost, the real cost based on performance characteristics and relative to the price of labour is in some cases declining and will likely continue to decline, thus stimulating applications. Estimates for Japan, for example, have put the ratio between the average yearly cost of labour in manufacturing to a playback robot as 1:10.5 in 1971 and 1:3.4 in 1979.

In addition to the direct cost of robots(10), it is necessary to take installation costs into account. These costs include accessories, maintenance costs(11), and indirect labour costs, which obviously need to be placed in context with savings in direct labour costs. For many robot applications it is also necessary to re-design the work station, and in some cases product re-design may be required. The technical skills and the expense which may be required in modifying the production structure could be an impediment to the adoption of industrial robots, particularly for SMEs. Some estimates put the cost of a complete robot system from about 40% to 150% in addition to the basic robot cost:

---

10. The lifetime of a robot is generally assumed as 8 years whereas the pay back period is usually completed in 2 to 4 years, often depending on whether a robot is used over one or two shifts.

11. Maintenance costs are usually estimated at about 10 per cent of the cost of the robot. For a firm which has not the internal capability and may therefore need to set up maintenance facilities, train labour, etc. maintenance costs can be high and may remain high unless multi-robot installations are envisaged.

46

DISTRIBUTION OF ROBOT COSTS

	Basic Robot Cost	Accessories	Installatio
Welding	55%	30%	15%
Handling	67%	22%	11%
Machining	45%	35%	20%
Machine Loading	55%	20%	25%
Painting	70%	24%	6%
Assembly	40%	35%	25%

Source: Le Nouvel Automatisme, Juin 1982

b) Labour Unions

Opposition by labour to existing robot installations has not been strong. Often applications have been encouraged by labour because these have been in health hazardous or monotonous workposts. Opposition may increase as robots are applied to assembly tasks, and despite the recognition by many labour groups that the use of robots will be necessary to maintain and/or increase a country's competitiveness, some concern exists as to the impact of robots on employment(12).

c) Financial Position of Firms

The weak financial situation of companies has been cited as one reason slowing down the application of industrial robots despite their apparent productivity(13). The present high cost of finance is adding to the difficulty of companies faced with a weak financial position. For small and medium enterprises the situation is often much more difficult. Even in spite of the apparent profitability of robots, lack of financial resources can be very much of a restraining factor. The interconnection between the need to increase international competitiveness, low profitability and a weak financial situation has to be recognised, particularly with respect to policy formulation.

---

12. In Japan, for example, concern for the future impact of robot applications has led to the commissioning of a two-year study by the Ministry of Labour to examine the social implications of these applications.
13. This slowdown in applications was apparent for many OECD countries in 1982.

d) Awareness

The lack of knowledge in the scope of robot applications is another factor which may slow down the rate of diffusion. The need for increased awareness is particularly relevant to SMEs. Recognising the need for awareness has led some countries, for example the United Kingdom, to implement specific policies to increase awareness and the opportunities offered by robots and CAD/CAM. Professional robot associations (e.g. the British Robot Association, Association Française de Robotique, Robot Institute of America, Japanese Industrial Robot Association) are playing an important role in increasing awareness at the general as well as technical level. In addition, these organisations serve as a focal point to bring together potential users, producers and policy makers.

e) Technological limitations

The technological capability of industrial robots, in particular the difficulty in identifying and picking up random parts(14), and the need for a high degree of repeatability and high absolute positional accuracy has restricted the use of robots to non-assembly work. At the experimental level, robots have been developed with high precision accuracy suitable for precision machine assembly. Expectations are that by the mid-1980s robots will be applied to assembly, based on the success of present trends in development efforts. Software design is expected to be a major constraint in development of robot applications, particularly for assembly work. Software needs include the requirement for greater machine integration (robot to robot and robot to other machines), the need for off-line programming (important in small-batch production), and the ability for robots to react to changes in external conditions and the environment.

f) Domestic production capability

Earlier, the importance some countries placed on the role of domestic producers in stimulating diffusion was noted. The relationship between the domestic production of a new technology and the domestic rate of diffusion of a new technology is not well documented in the economic literature. A recent study on textile machinery concluded that "commercial R&D which takes place within a nation's boundaries can benefit domestic diffusion in a way not achieved by the importation or imitation of foreign

----

14. The problem of sensory ability for robots, and other automation systems, is probably the key area at present under consideration by researchers.

48

technology"(15). But this analysis also concluded that the case for textile machinery cannot be generalised for all capital goods. Intuitively, in a country with a dynamic and highly competitive producing industry where firms are vying for market shares as well as long-term growth, it can be expected that diffusion rates will be significantly higher than in a country relying mostly on imported technology. The Japanese situation can be cited as an example in this regard. At the same time, the dynamism and acceptance of a new technology by using firms is also of importance.

Domestic capacity constraints have also acted as a brake on domestic diffusion in some countries. Countries with a fairly high rate of domestic robot production and high rate of domestic demand have, until recent years, concentrated on satisfying domestic demand, consequently export sales have been low. Countries with small domestic production capability have not always been in a position to satisfy domestic requirements adequately.

g) Engineering manpower shortages

A factor inhibiting demand has been in some countries (for example in Canada and Australia) a shortage of engineers and technicians with experience in robot applications (both in terms of hardware and software). In such cases policies which assist firms, particularly small firms, in defraying consulting engineering expenses may be appropriate. In the longer-term, as recommended in a recent report on robotics by the Australian Science and Technology Council, countries may need to reallocate resources within their higher education system in order to ensure that the number of graduates with appropriate skills in computer-integrated manufacturing systems is sufficient.

*
* *

It is difficult to isolate which particular factor has played or will play a more important role in stimulating demand for robots. This often depends on the area of application and the particular circumstances in different countries. In most cases, however, the net result, because of cost savings and/or increased productivity, is that investment in robots is economically profitable and thus takes place. The net effect of the above factors has

---

15. Benvignati, A. M., The Relationship between the Origin and Diffusion of Industrial Innovation, Economica 49, page 319.

led to a significant growth of robot installations, which accelerated in the late 1970s and early 1980s. The world population of industrial robots has grown from about 1,000 in 1970 to approximately 22,000 by 1981 (a growth rate of 32 per cent per year). Growth has differed significantly between countries (Table III.9). At present, approximately two-fifths of the world stock of industrial robots is in Japan and one-fifth in the United States. Germany and Sweden have the highest population of robots in Western Europe, a share of the world stock of approximately 10 per cent and 8 per cent respectively. Japan's share in the total stock has been increasing since 1978, while the relative share of the United States is declining because of its slower rate of applications. The United Kingdom and France have both been slow in adopting robots in the 1970s, but have shown a significant increase in the last few years(16). Sweden, on the other hand, which showed

Table III.9

ESTIMATES OF THE POPULATION OF ROBOTS,
VARIOUS COUNTRIES(1)

	1974	1978	1980	1981	1982	Average Annual Growth	
						1981-82	1974-82
Japan	1500	3000	6000	9500	13000	37%	31%
United States	1200	2500	3500	4500	6250	39	23
Sweden	85	800	1133	1700	1300 (2)	-	41
Germany	130	450	1200	2300	3500	52	51
Italy	90	n.a.	400	450	790	56	31
United Kingdom	50	125	371	713	1152	62	48
France	30	n.a.	580	790	950	20	54
Netherlands	3	4	49	62	..	-	-

Sources: 1974 and 1978 - various sources. 1980: British Robot Association; France Diebold. 1981: British Robot Association; France - BIPE. 1982: British Robot Association. 1974-81: Netherlands - Stichting Toekomstbeeld der Techniek

Note: Data based on a narrow definition of robots.

1. Accurate data on the world population are not available, but the following estimates are indicative: 1980 - 13,700; 1981 - 22,000; 1982 - 31,000.
2. Data revised downward as a result of definitional changes.

---

16. An aspect of demand which could be relevant here is that during slow growth periods firms place more importance on rationalising and cost-saving investments, increasing the attractiveness of robots.

significant growth in robot applications during the mid-1970s, has been experiencing a slowdown in the rate of applications.

One measure of the rate of robotisation is the number of robots used per worker in manufacturing (Table III.10). Sweden's rate of penetration in this regard is significantly high, stimulated perhaps by the relatively high unit labour costs and the need by firms which are highly export oriented to retain their competitiveness, but also by a government policy which, since the early 1970s, has encouraged the adoption of new automation technologies. Such a high ratio of robots to manufacturing employment is likely to be attained only by Japan, among the major OECD industrialised countries, before the mid-1980s, unless these countries experience a significantly high rate of applications over the next few years.

Table III.10

ROBOTS PER 10,000 EMPLOYED IN MANUFACTURING

	1974	1978	1980	1981
Japan	1.9	4.2	8.3	13.0
United States	0.8	2.1	3.1	4.0
Sweden	1.3	13.2	18.7	29.9
Germany	0.4	0.9	2.3	4.6
United Kingdom	0.1	0.2	0.6	1.2
France	0.1	0.2	1.1	1.9

Source: Employment data: Indicators of Industrial Activity, OECD

Forecasts on the rate of growth for robots range from around 30-35 per cent per year for the world over the next decade. Although high, this growth rate is from a low stock level. Demand can be expected to accelerate up to 1985 as workposts where robot technology is well developed, and the use of robots proven, are converted rapidly to the use of flexible automation, and slowing down thereafter as more complex tasks, particularly in assembly, are robotised. A growth rate from 1981 of 30-35 per cent would imply a stock of industrial robots of between 230-330,000 by 1990. Demand in the 1980-1990 period will be mostly new demand, while replacement demand will become more significant after 1990.

Various forecasts are available on trends in the future demand for industrial robot installations in some countries:

Japan	1980-1985:	35-40%	(Bache, Halsey Stuart Shields)
		30%	(JIRA)
		10-20%	(CREI-Lettre 2000)
United States	1980-1985:	35%	(Wall Street brokers)
		26%	(CREI-Lettre 2000)
	1980-1995:	36%	(Predicast)
Sweden	1979-1984:	20%	(Swedish Computers and Electronics Commission)
	1985-1990:	17-26%	(ibid)
	1981-1986:	26%	(Creative Strategies International)
France	1980-1985:	27-35%	(Diebold)
	1985-1990:	21-24%	(Diebold)
	1981-1986:	49%	(Creative Strategies International)
United Kingdom	1981-1986	58%	(Creative Strategies International)
West Germany	1981-1986	46%	(Creative Strategies International)
Italy	1981-1986	56%	(Creative Strategies International)
Australia	1982-1990	30%	(Australian Science and Technology Council)

In general, it is expected that the two major markets will remain as Japan, reaching about US$700 million (using a narrow definition of robots) in 1990(17), and the United States, at about US$1.5 to US$2 billion in 1990. Although this implies a rapid growth, the industry will still be fairly small (for example present sales of machine tools in the United States are about US$5 billion and US$1.2 billion for NC tools). The Western European market is expected to attain yearly sales of US$500-700 million by the mid-1980s, possibly reaching US$1 billion by the 1990s.

.

---

17. Some private analysts have estimated a significantly higher annual sales figure by 1990.

Chapter IV

THE IMPACT OF INDUSTRIAL ROBOTS ON INDUSTRY

1. INTRODUCTION

The assessment of the impact of a new technology is invariably difficult. Many studies which attempt such an assessment rely on direct questionnaires/surveys of a sample of user firms and undertake an assessment retroactively; that is, once the new technology has been well established. The difficulties raised when examining the future impact of a new technology have limited many studies to a discussion in a general way of some of the salient issues. Thus:

"...a general opinion of many National Commissions [assessing the impact of microelectronics is] that it is an almost impossible task to be definitive in quantitative terms about the matters under discussion, and it is therefore safer to resort to qualitative rather than quantitative conclusions."(1)

This is true as well in assessing the impact of robots on industry.

In the case of robots there is the need to differentiate between the impact of robots as a whole on industry and the impact of specific applications in various individual firms, workposts or industries. The former case has to rely on a more general, analytical approach, while the latter type of applications are best looked at from the point of view of individual case studies. However, even such cases are limited by data constraints, for example data on how many workers will be displaced in the workposts where robots are being applied, what the impact on

---

1. Stoneman, P.; Blattner, N.; Pastré O.: Information Technology, Productivity and Employment, OECD, ICCP Publication Series No. 7, Microelectronics, Robotics and Jobs, 1983.

output is and what the penetration rate(2) will be. There is an added difficulty in that, despite the substantial increase in productivity of a particular application, it may be difficult to trace this increase through a plant as a whole.

2. THE INDUSTRIAL IMPACT OF ROBOTS

The impact of robots on industry will largely be seen in the following ways:

i) higher productivity; productivity increases can be expected in the technical sense (e.g. number of cars painted per hour, etc.), but more importantly in the economic sense, that is, in reducing unit costs. The productivity impact depends basically on the "statistical weight" a certain workpost, production line, etc. has in the total production process(3), and will obviously increase where there are intense applications (e.g. assembly) or where a particular work function is crucial to the production process. Thus in the small-batch metalworking industry where robots may play an important role, only a few applications may be required to raise firm productivity significantly, but since this industry has a relatively small role in total manufacturing, the impact on total manufacturing productivity may be marginal. In the case of automobiles, electrical/electronics industries, and machine manufacturing industries, their greater weight in total manufacturing would result in greater total productivity increases if the introduction of robots raised productivity in these industries.

ii) improved product quality; this results from greater product consistency and as a result of a reduction in human error. In many workposts,

---

2. The penetration rate refers to the number of robot installations of a certain type as a percentage of workplaces of that type in the manufacturing sector.
3. A hypothetical example can illustrate this: if a particular work task which is robotized accounts for 20 per cent of plant activity and the productivity increase in that task resulting from the introduction of robots is 30 per cent, then total plant activity may increase by 6 per cent. If all the industry is robotized and that industry contributes 10 per cent to manufacturing output then the total impact on manufacturing productivity would be 0.6 per cent.

a manual operative is usually faster than a
robot. However, a robot maintains a constant
speed and therefore tends over an 8-hour period
to do better than manual operators, whose speed
and concentration are reduced over the latter
part of a working shift(4). In terms of output/
rejects ratio and output/quality ratio, robots
tend to surpass manual operators. It should be
noted in this context that quality improvements
would not per se be reflected in conventional
statistics measuring productivity.

iii) the social impact; labour displacement, a change
in occupational structure, and a change in work
methods and organisation will all occur as a
result of the introduction of robots and other
flexible automation (see section below).

iv) the structural impact on manufacturing; robots
and other elements of CAM may affect the structure of industries through their impact on relative competitiveness and international trade
flows. However, the direct structural impact of
this technology is not expected to be significant.

v) manufacturing concepts; the increasing use of
robots in industry and their integration with
other machines and the increasing interface of
robots with NC machine tools, CAD and CAM systems, are all leading to changes in manufacturing concepts and procedures. These changes cannot be attributed to robots alone, but their
role is important.

vi) price effects; industries with a greater scope
for robotisation may through improved productivity show better price performance stimulating demand for their products, and consequently
positively affecting growth.

*
*   *

However, there are limitations to the impact of
robots in particular work stations, such as:

i) the volume of throughput; this may increase in a
particular work station (for example, spot-welds

_____

4. The average productivity and marginal productivity
of an industrial robot are therefore equal, and marginal
productivity of a robot is greater than that of manual
labour.

55

per hour, etc.), but such an impact may not be transferred throughout the whole plant if bottlenecks exist or other machinery/labour cannot speed up their throughput, so that although productivity has increased it is localised to a particular work station. Thus, the efficiency of an individual piece of capital equipment is not in itself sufficient to increase total plant or industry productivity.

ii) cost impact or savings; there will obviously be an impact on the final product in terms of cost, but while the savings for the particular work station may be large, the savings impact on the final product would depend on the relative importance that the particular work station plays in the total production process. However, it is important to stress that the cost savings resulting from the introduction of robots are not only in terms of labour, but include (a) material costs as a result of a lower ratio of rejects and better machining tolerances; (b) inventory costs, as a result of lower lead times; (c) capital costs, since some robot lines require less floor space than conventional lines.

iii) the number of jobs; the number employed in a particular work station may be reduced but overall employment in the plant may not decline as a result of labour reallocation or because the increased use of skilled labour compensates for displaced, unskilled labour. The foregoing assumes that the plant increasing productivity through the introduction of robots benefits from increased output as a result of increased competitivity. With output for the plant remaining constant, some fall in labour usage can be expected if there is to be some productivity increase as a result of introducing robots.

Investment in industrial robots is rationalising rather than capacity expansion investment since the primary aim is to increase productivity per se and as such robots tend to be capital-intensive as regards the capital-labour ratio. In terms of the capital-output ratio, however, it is not easy to determine whether industrial robots raise capital intensiveness or not, i.e. some robot applications represent substitution of new forms of capital for old forms of capital. The functions to which they are being applied and the required ancillary equipment need to be examined in this regard.

An important aspect of robots is their flexibility allowing desynchronisation in output, as well as in the ability to link and effect transfers between various machines and thus assist in reducing machine waiting time. As such, robots help increase productivity of other capital goods and lead to a reduction in capital costs per unit of output. This impact on capital productivity is an important consideration(5).

The impact of industrial robots will not be the same across all industrial sectors. Many of the process industries are already highly automated (chemicals, petroleum refinery, pulp and paper, food processing, etc.). While there may be scope for further or more efficient automated systems in these industries, particularly through increased use of computer-aided manufacturing systems, this is unlikely to be achieved through the use of robots per se except in some workposts (e.g. palletizing).

The industries where robots have penetrated at present tend to be highly mechanised. These industries include the manufacture of fabricated metal products, the manufacture of machinery (in particular metal working machinery and special industrial machinery), the manufacture of electrical machinery, and the transport equipment industries (especially motor vehicles and, to a lesser extent, aircraft). In some of these industries, substantial rationalisation and adjustments in the production structure are taking place and, while robots form part of this process, they do not necesarily form the major part. In this regard, for example, the impact on labour displacement by robots has been minor in some industries compared to labour shedding through the general streamlining of production.

Many of the industries which are active users of industrial robots (e.g. automobiles, electric-electronic machinery, precision machinery and metal-working industries) have applied robots to areas so as to eliminate danger to workers, areas which are health hazardous or dirty, and to eliminate monotonous work. This is not to say that productivity aspects were not a consideration as well. On the contrary, payback calculations would invariably play a part in the investment decision. Future motivations in introducing robots are, however, more likely to emphasize the requirement to increase productivity, that is, the desire to reduce costs and reduce interruptions and restrictions on production throughput. The emphasis on productivity has become particularly important given the increased recognition that industry's fundamental problem is the decline and stagnation of productivity growth.

_____

5. The role of computer-aided manufacturing in integrating machinery as well as monitoring machinery will perhaps result in a better knowledge by plant managers of the productivity of capital equipment.

The future impact of industrial robots will depend not only on their expected rate of diffusion, but particularly on the successful development of intelligent robots which can be applied to assembly tasks. As previously noted, the emphasis of present research and development work is mainly on development of pattern recognition systems and other sensory devices in order to develop an industrial robot which would be effective in assembly work. Research on sensor systems has been underway since the early 1960s with significant progress being made. Nevertheless, the complexity required in analysing random parts in manufacturing would indicate that successful commercialisation of industrial robots for use in complex assembly is unlikely before the late 1980s. However, for more simple assembly operations, the successful introduction of robots can be expected by the mid-1980s. A Delphi forecast for the United States undertaken by the American Society of Manufacturing Engineers (in conjunction with the University of Michigan) predicts the following trends:

1. By 1982 5 per cent of all assembly systems will use robotic technology;
2. By 1985 20 per cent of labour in final assembly of automobiles will be replaced by robots and automated systems;
3. By 1987 15 per cent of all assembly systems will use robot technology;
4. By 1988 50 per cent of labour in small-component assembling will be replaced by automation;
5) By 1990 development of sensory techniques will enable robots to approximate human capability in assembly.

While possibly optimistic, these forecasts are indicative of the expected changes which will affect manufacturing industries and particularly the assembly industries over the next decade. Other estimates indicate a similar emphasis in distribution in favour of assembly intensive industries (Table IV.1). For example, in the United States, which is expected to have a higher rate of robot penetration than France by 1990, light manufacturing will be the main market for robots followed by automobiles. Both industrial sectors are assembly-intensive, as is the electrical/electronic sector which is expected to be the third largest market for robots.

On a functional basis, the forecasts of the expected distribution of robots (Table IV.2) indicate a fairly rapid increase and then a relative levelling or decline in the medium term in the application of robots in posts where penetration rates are expected to reach saturation levels by the mid-1980s (e.g. spot welding and painting). In the longer term, a fairly rapid increase in the use of robots in assembly and control functions can be expected. In assembly, for example, the forecast distribution of robots is very similar between Japan, the United States

Table IV.1

Percentage Distribution of Robot Sales by Industry		Percentage Distribution of Robot Units by Industry	
United States	1990	France	1990
Light Manufacturing	39%	Automobile	30%
Automobile	23%	Metal Working	20%
Electric/Electronic	14%	Electronics	18%
Foundries	11%	Mechanical Engineering	15%
Heavy		Material ceramic,	
Manufacturing	9%	transformation	4%
Aerospace	4%	Plastic Material	4%

Source: Bache, Halsey, Stuart Shields

Source: Diebold Consultants

and Germany, where it is estimated to reach 20 per cent of
the total robot stock in each country. Other estimates of
the penetration of robots in industry indicate a much
slower rate (Table IV.3, section 6 below). Thus estimates
for the UK indicate that even in workposts where applications of robots are proven (e.g. spot welding, injection
moulding) penetration rates by the end of the decade are
only half of the potential. Areas where the number of
potential applications are high, such as assembly, are
estimated as having very low penetration rates by the end
of the decade.

It is not clear whether the impact of industrial
robots will be concentrated in those industries which,
because of low productivity growth, could perhaps benefit
the most from the introduction of flexible automation. In
some countries, such as the United Kingdom, industries
adopting industrial robots are those with low productivity
growth. These industries which include mechanical
engineering, aerospace, metal goods and instrument
engineering, were identified by the Advisory Council for
Applied Research and Development in the United Kingdom as
key industries which make a large contribution to
industrial output but have under-average productivity and
would thus benefit from increased automation. Yet in
other countries these industries already have relatively
high productivity growth and are also adopting robots.
Thus, one cannot necessarily analyse the industrial
pattern of diffusion just on the basis of the requirement
for productivity increases. Rather this pattern has
depended on the suitability of first generation robots to
the various sectors.

Industries where industrial robots can also be important are those where market saturation has been, or is
being, reached and therefore tend to be very price-
sensitive and require production rationalisation in order

Table IV.2

FORECAST DIFFUSION OF INDUSTRIAL ROBOTS BY FUNCTION

JAPAN
(% distribution by value)

	1985	1990
Spot Welding	10	8
Arc Welding	11	10
Painting	5	5
Surface Treatment	1	1
Machining	14	13
Assembly	17	22
Control	8	10
Plastic Moulding	4	3
Pressure Casting	2	2
Casting	2	1
Heat Treatment	3	2
Forging	2	1
Metal Cutting and Stamping	4	5
Other	17	17

UNITED STATES		GERMANY		
(% distribution by number)		(% distribution by number)		
	1991		1985	1990
Spot Welding	13	Spot Welding	20	11
Arc Welding	10	Arc Welding	3	2
Loading/Unloading Machines	15	Machine Tools	23	26
Handling	20	Press Tools	21	21
Painting	5	Surface Treatment	7	4
Machining	7	Grinding, Drilling	6	4
Assembly	23	Assembly	10	21
Other	7	Other	10	11

Sources: Japan - JIRA; USA - Le Nouvel Automatisme, June 1982; Germany - Volkholz, V; in OECD, ICCP Publication Series No. 7, op. cit.

to adjust to lower market growth (automobiles, household appliance industries). These industries also require great production flexibility in order to adjust to changing consumer requirements and technological availabilities. The textile, clothing and leather industries have also been cited as a likely area for future employment of industrial robots because of the low productivity of the industry and the strong competition from LDCs. In Sweden and Japan, for example, significant research efforts are under way to automate clothing production on the basis of robots and CAD/CAM systems. Important development work is also under way in Japan to introduce robots in the ship-building industry.

In summary, for the medium term, that is to the mid-1980s, it can be expected that the use of robots will intensify in the industries which are at present the main users and in those workposts which are now being robot-ised. A fairly high penetration rate of robots will occur in such posts as spot welding, surface treatment (especi-ally painting), machining, injection moulding. After the mid-1980s a fairly rapid growth in the use of assembly robots can be expected, but it is nevertheless unlikely that a high penetration rate will be achieved before the end of the 1990s, especially for complex assembly. How-ever, despite estimates for high growth in the stock of robots in various countries the size of this stock will still remain relatively small by 1990 (see Table IV.4), in comparison with the stock level if all potential workposts in manufacturing, including assembly posts, were filled. Obviously the size of the stock will have implications for the total productivity impact of robots.

3. THE IMPACT ON ASSEMBLY-INTENSIVE INDUSTRIES

Assembly problems have usually been solved by a high division of labour and therefore specialisation in the tasks performed allowing for the use of unskilled labour. This simplification of work procedures will facilitate the introduction of flexible automated systems. The impor-tance of assembly employment in some key industries is apparent from the following table (the data are based on the situation in Italy, but are relevant for most indus-trialised countries):

Industry	% of Labour Force Employed in Assembly
Automobiles	32
Agricultural Machinery	31
Domestic Appliances	23
Light Electromechanical Industry	43

Source: A. d'Auria, Automation in Batch Assembly, Seminar on Innovation in Engineering Industries, Turin, 1980, UN, Economic Commission for Europe.

It is also fairly clear that automation has not sig-nificantly affected assembly industries so that from 77 per cent to 86 per cent of assembly operations are presently performed manually (the data refer to Canada, but are generally applicable to most industrialised countries):

Percent of Assembly Performed by Method

	Subassembly	Final Assembly
Manual Bench	46%	41%
Manual Progressive Line	31	45
Semi-automatic Machine	17	10
Automatic Machine	6	4

Source: National Research Council Canada: Robots in the Canadian Manufacturing Industries.

There is thus substantial scope for the introduction of robots in assembly-intensive industries once technological obstacles are overcome. Even assembly industries already highly automated can also benefit from the use of flexible automated systems since their existing equipment is usually expensive and highly dedicated as well as limited to non-complex tasks(6).

Automation and labour-saving in the assembly phase of production will become one of the most crucial areas in boosting productivity. Many companies which are now investing in robots are doing so as a defensive strategy in order to gain experience before the next generation of "smarter" sensory models are available, in particular models with vision and tactile capabilities.

In addition, the rapid progress towards the automated factory and automated assembly processes could provide a substantial competitive advantage to countries adopting this technology - both in terms of productivity and quality - which could not be emulated by assembly units with a human workforce. Therefore, enterprises (and countries) which gain the necessary experience in the use of robots now will be able to adopt quickly and easily the expected future generations of industrial robots.

The future impact on assembly industries can to some extent be illustrated by the plans of larger companies,

---

6. The question of the optimal production volume for which it is economical to introduce robots needs to be considered. Some studies have indicated that the optimal use of robots for assembly is for approximately 100,000 units, i.e. an intermediary volume between manual assembly and assembly based on rigid automation. However, when product changes are foreseen then the long-term unit costs would be cheaper if robots were used for smaller volume runs.

many of which play a major role in the manufacturing sector and are presently reviewing their manufacturing process with the aim of introducing industrial robots:

i) In the United States, General Electric is expected to launch a substantial automation programme which, once in place, could eliminate up
to 2,000 blue-collar workers. General Electric
has already installed robots in its consumer
appliance plants, particularly for fairly simple
tasks. In 1981, GE was planning to spend $5.1
million on robots and save $2.6 million in
labour and material costs (i.e. a two-year pay-
back period). Such savings provide a substantial incentive for further applications to the
company as well as having a significant labour
displacment impact. In addition to productivity
increases and labour-saving, GE has found that
robots also help substantially in material
saving.

ii) Hitachi in Japan, producing a similar range of
output to General Electric, undertook an extensive analysis of the production process in its
plants which determined that the share of assembly work takes 35 per cent of total working
hours. Automating assembly work, it is estimated, could reduce those employed (in assembly)
by 30 per cent and have a significant impact on
productivity. Consequently, Hitachi has embarked on an extensive R&D programme for an "intelligent" assembly robot. The expected time
frame for conversion is for the mid-1980s, but
is obviously dependent on success in present
development efforts.

iii) A leading Japanese company, Fanuc Limited, is
operating a plant producing numerically controlled machine tools and robots by using NC machine
tools and robots. In the first year of operation the plant will employ 100 workers to produce 100 robots per month and a large number of
machine tools - the number of workers is about
20 per cent of what a normal plant would require.

iv) One of the more ambitious projects in automation
is in Japan where a project to build the world's
first fully-automated factory for turning out
small batches of engineering components is under way. The test plant is scheduled for operation by 1983 although a longer period will be
required to complete the construction of a fully
automated factory. The factory is designed to
produce different metal parts and automatically
assemble complicated products. Robots integrated with NC machine tools will play a crucial
role in the factory.

v) Pioneer in Japan (audio, consumer electronics) plans to use flexible manufacturing systems in assembly and has estimated that by 1985 90 per cent of electrical components will be assembled by robots.

vi) A simple Fanuc assembly robot used in the assembly of servo motors can assemble over 80 motors in a 7-hour period whereas during this period a worker can only assemble 30 motors.

It needs to be stressed that the impact on assembly industries will, to a large extent, depend on the degree to which robotics can emulate the adaptability of humans in manipulation, in sight, in thinking and in flexibility. The need is for robots with high speed, flexible gripper systems and sensor ability. Furthermore, the level or degree of automation will be determined by economics more than by technical possibilities and, in particular, the pay-back period will be important for investment appraisal. Such new technical capabilities are likely to be phased-in over a long period, so that the impact will be evolutionary. However, it is clear that successful automation of electronic and electrical engineering goods would have a considerable impact in terms of productivity and labour-saving throughout the electrical engineering sector of OECD countries. A similar impact can be expected in the mechanical engineering industry, in particular the successful integration of NC machine tools and industrial robots will have an impact on the machine tool industry in terms of employment displacement and productivity increases. In addition, it cannot be expected that any significant number of jobs will be created by the growth of the industrial robot industry when the introduction of flexible automation becomes widely used in the producing industry.

4. THE IMPACT ON BATCH PRODUCTION

Approximately 50 per cent of manufacturing uses batch production methods, varying of course from country to country and according to sector(7). Batch production, although not necessarily restricted to small and medium enterprises (SMEs), is of particular relevance to sectors such as metalworking, engineering, plastic forming, etc. where SMEs are often dominant(8). Given the economic

---

7. In the United Kingdom, for example, about 40 per cent of all engineering products are produced in batches of 50 or less. In the US batch manufacturing represents about 35 per cent of manufacturing output.
8. See Table 4 in Innovation in Small and Medium Firms, Background Reports, Part 1 (OECD, 1982).

importance of SMEs in terms of employment and value-added in the manufacturing sector(9), and the wide use of batch production in SMEs, any significant productivity impact by robots and other flexible automation would have important consequences. However, the availability of robots, CAD/CAM, etc., does not imply that automation in discrete engineering is easy. Management skills and expertise are often very important in this context and can be the main barrier against the diffusion of new forms of automation.

The potential benefits which robots can provide to batch production techniques are significant. The use of robots in piece handling, tool changing, etc., in machining centres increases the speed of throughput substantially, reduces inventory requirements and results in a higher total output and reduced unit fixed costs. In particular automatic tool changing and adjustments reduce considerably the amount of machine idle time which has been a prime factor in maintaining relatively low machine productivity. In batch manufacturing machining sometimes takes up only 5 per cent of the time, and the remaining time is for waiting, set-up, assembly, etc. Automatic tool changing and self-adjustment also allows for machining with unsupervised night shifts. The ability to reduce waiting time facilitates an increase in volume and in distributing capital and other fixed costs over a larger volume. An important aspect of these factors is the impact on increasing capital productivity as well as labour productivity. The flexibility of robots (programmability) also helps make small production runs more economical by allowing overheads to be spread over a wider range of products. The use of flexible automation can thus help reduce the advantage enjoyed by products where high volume production is possible(10). This ability to produce a variety of parts, easily and economically, reduces the dependence by firms on a small market segment. Finally, the use of robots can help reduce the demand for skilled machinists who in many OECD countries are in short supply (e.g. Japan, UK).

The flexibility provided to batch manufacturing by new forms of automation is not solely due to industrial robots. The increased flexibility is a result of the increasing use of numerically controlled machine tools, the ability to integrate machine tools into machining centres, the integration of robots into systems and direct numerical control of manufacturing systems.

---

9. SMEs account for between 45 to 70 per cent of industrial employment in OECD Member countries, ibid Table 1.
10. This does not necessarily mean that unit costs under batch manufacturing will become equivalent to those for mass produced parts.

Of particular relevance to batch production is the use of flexible manufacturing (FM) systems which are being used to produce a variety of parts particularly in the machine tool industry as well as in other machinery indus- tries. FM systems involve the integration of various types of machine tools, usually by robots which service the tool and handle pieces and which are directed by a central computer. FM systems are usually used in machin- ing parts, for assembly into larger parts, machining parts which require similar machining processes or parts which are similar in shape(11). This integration of automated flexible machinery has led to the increasing use of machining cells which can produce a variety of parts. This ability to produce a variety of parts may mean that batch manufacturing could be integrated into a plant by larger firms and thus could possibly affect small and medium enterprises.

The productivity gains from FM systems can be signi- ficant so that, in spite of initial high capital costs, the expected payback period is short. An often cited example is the productivity increase of a Japanese company (Yamazaki Machinery Works Limited) where, as a result of the introduction of a flexible manufacturing system, the plant takes one month to produce 74 different types of parts (1,200 pieces). A conventional factory would nor- mally require 3 months.

One aspect encouraging the application of robots, FMSs, etc., in SMEs is the desire for many companies to evolve from being subcontractors into primary suppliers, at the same time maintaining their status as a SME. Flexible automation provides these firms with the ability to produce a variety of products with limited production runs, which is required if they wish to be primary sup- pliers.

Some examples where flexible automation has increased productivity for SMEs include:

   i) a Japanese producer of plastic VTR cassette
      cases introduced flexible plastic extruding
      machines serviced by industrial robots resulting
      in a doubling of output without increasing the
      labour force;

  ii) an automobile seat frame producer in Japan auto-
      mated the welding process using robots and esti-
      mated that labour productivity increased four-
      fold and quality improved;

---

11. FM systems are already being used to produce cut- ting tools, machine tool parts, engine parts, robot parts, etc. FM systems are expected to be used in plastic injec- tion moulding and precision moulding in the near future.

iii) a French producer of kitchen cabinets has robotised its painting line to varnish cabinet
doors. The result has been a 50 per cent saving
in varnish and the treatment of 12,000 doors per
day by six operators using two robots in place
of 100 operators;

iv) a Japanese manufacturer of diesel engines and
other machinery introduced flexible manufacturing systems which reduced labour requirements
from 31 to 4 and allowed for continuous operation at night without labour whereas previously
the plant worked only on one shift;

v) a Japanese producer of textile machinery parts,
machine tool parts, etc. introduced FMSs and, as
a result, reduced production costs from between
33-50 per cent and has started operating unmanned night shifts;

vi) an Italian forge shop has used robots to reduce
press forge operators (a reduction of 33 per
cent of the forge shop workforce) and increased
output from 400 to 1000 pieces per hour. Robots
are used for loading and unloading furnaces,
forging presses and conveyors.

Many SMEs are in a position to purchase flexible
automation such as industrial robots because of their
relative cost performance. In the case of a robot its
cost is equivalent to about two years' pay of a skilled
worker. Despite the apparent profitability in the use of
robots, many small and medium enterprises are reluctant to
invest in such systems because of the inability to obtain
financing, lack of awareness or lack of expertise.
Government schemes which help increase awareness, provide
consulting engineering assistance (e.g. as in the United
Kingdom) or the leasing scheme as used in Japan can be
useful in stimulating diffusion of robots in small and
medium enterprises.

In order to introduce a robot onto the shop floor, a
SME may have to reorganise its production operations, in-
curring greater costs than the initial robot costs. The
firm may also find that while efficiency increases with
the introduction of a robot or robots, maximisation of
efficiency can only be achieved through "total automation"
rather than partial, thus requiring investment in, for
example, other forms of flexible automation (CAD/CAM systems, automated carriers, etc.). In such cases, firms may
prefer to defer investment in robots until they are in a
position to invest in more comprehensive systems of automation.

5. THE IMPACT ON THE AUTOMOBILE INDUSTRY

Introduction

The evolution of automobile production methods over the years has been towards standardization and the reliance on the efficiency of mass-production methods. The trend in automation of production in the pre-war years, as well as significant improvements in the 1950s and early 1960s, enabling mass production techniques to be applied to automobile production, led to the maintenance of relatively low product prices and to a product which became widely available as a mass consumer good.

Automation which affected the industry took place particularly in the engine plant, in the manufacture of body parts, whereas in final vehicle assembly the impact of automation has been much less. Some aspects of these changes should be noted. Firstly, they led to a high growth in productivity. Secondly, there was considerable labour shedding, particularly during the early 1960s in areas such as engine assembly. Thirdly, standardization and mass production led to the use of highly dedicated capital goods.

The latter meant that production became relatively inflexible. For example, small design changes required heavy investment for retooling and often plant shutdown for several months. Thus, there was a tendency to minimise such changes, particularly in highly automated facilities such as an engine plant. As a consequence, it could be argued(12) that, despite the high volume throughput and higher productivity of dedicated facilities, in the longer run technical innovation was retarded. As a consequence, productivity growth also declined.

The decline in productivity growth, combined with an increasing cost-squeeze as a result of high and growing hourly labour costs, shifted producer emphasis towards increasing plant efficiency. Thus, in the late 1960s, a slow trend began towards the introduction of more automated production methods in the automobile assembly plant itself which, nevertheless, remained the most labour-intensive area. For example, in 1969 the first industrial robots were introduced in welding and painting in the assembly line. However, partly because of technological constraints and partly because vehicle sales remained buoyant, the emphasis by many producers on productivity improvements remained fairly marginal. Recent exogeneous shocks to the industry, particularly fuel price increases,

---

12. See, for example, Abernathy, W. J., "The Productivity Dilemma: Roadblock to Innovation in the Automobile Industry", Johns Hopkins University Press, Baltimore 1978.

the slowdown in demand growth and the erosion of market shares as a result of intensified international competition, have placed pressure on automobile producers to move away from dedicated machinery, suitable only for non-variable mass produced products, towards flexible production methods.

The reduced reliance by automobile producers on specialised machinery in favour of more flexible automation, which is offered by industrial robots and other CAD/CAM equipment, is viewed as necessary because:

   i) there is a need to reduce costs and increase productivity in the face of international competition;
  ii) anticipated rapid technological change, both product and process, can be integrated more easily;
 iii) the uncertainty in demand, both in volume, as well as product type, can be handled more easily.

On the cost side, rising labour costs combined with declining labour productivity have usually been cited by producers as the main factor in reducing their ability to compete effectively. Factors such as increased product quality and reliability are also playing a major role in the strategy of producers to improve competitiveness.

On the technological side, the existing product offered by automobile producers and market requirements are not in equilibrium given consumer emphasis on energy-saving, safety and reliability. This implies that extensive product innovation over the next decade is required, and may also be necessary, in order to expand markets by increasing replacement rates by offering a new, more innovative product. If this does not occur, the approach to automobile demand saturation in many countries would indicate even lower growth rates in the future as compared to the present rates(13).

Impact on Automobile Production Structure and Costs

The prospect of declining rates of growth in demand vis-à-vis existing production capacity levels would indicate that in the long term the automobile industry may face a period of mergers and capacity rationalisation. This could arise since it would become more difficult for some producers to maintain an economic level of output as more efficient producers increase their market share. A tendency towards rationalisation will be reinforced by

13. See, for example, Long Term Perspectives of the World Automobile Industry, Chapter II, OECD 1983.

flexible automation techniques which will enable production capacity to meet changing demand levels, changing model styles and technological requirements fairly easily.

Standardization and efficiency through economies of scale have become the norm in the automobile industry. As a corollary the production process became more rigid as the level of mechanisation increased and production became dependent on specialised equipment. The importance of the present developments in flexible automation is that such equipment allows producers to move away from rigid mass production methods allowing product changes to be incorporated without the need to scrap expensive dedicated machinery(14).

Two aspects of the increased flexibility in production need noting. Firstly, the concept of minimum efficient scale for an automobile assembly plant as well as for such components as an engine or transmission plant is changing. For example, a plant producing 400,000 vehicles a year was considered as an optimum size whereas, with increased automation, manufacturers are considering a plant with a throughput of about 250,000 vehicles as optimum. The reduced minimum efficient scale could, based on the expectation of continued slow growth in demand in the medium term, allow many of the larger firms to rationalise capacity by adopting flexible automation without altering unit costs and profits(15). Under a production structure where 400,000 units a year was optimum, smaller firms operating at this scale would suffer if, as a result of eroding market shares, they had to adjust capacity. Rationalisation based on flexible automation would also allow such firms to adjust successfully(16).

---

14. A recent example is an attempt by the Ford Motor Company to lower unit costs through large-scale production in a new engine-block factory which turned out to be too inflexible to be efficient. The inability to convert the casting plant to build smaller engines efficiently led to its closure.
15. A recent report by an investment analyst in the United States has indicated that break-even points by the major producers have been significantly reduced: GM from 6 million to 4.8 million units, Ford from 4.2 million to 3 million units and Chrysler from 2.2 million to 1.3 million units. This change in break-even levels is indicative of the extent to which producers are reducing their overhead costs in conjunction with efforts to increase manufacturing efficiency.
16. While minimum efficient scale for a plant may change, economies of scale for automobile production as a whole will remain important, particularly as regards standardized components, which are used in a large variety of models (e.g. engines, transmissions), and the need to

(Continued on next page.)

The second aspect of increased flexibility in production relates to the ability to incorporate product changes easily. This will allow technological changes to be adopted more easily than in the past since the production equipment used is less dedicated. In the longer term, therefore, production flexibility will help stimulate technological change and facilitate innovation. In addition, exogenous shocks such as fuel price increases, environmental concerns, etc., which have provided an impetus for technical change in automobiles can be incorporated much more easily and at a lower cost. The net result of increased production flexibility in the long run could be a reduction in investment requirements in the industry by eliminating the need for constant retooling to meet product and component changes. This reduction in investment requirements could, besides cost savings, have an impact on the traditional linkages between the automobile producers and the machine tool industry.

However, in the short to medium term, investment requirements for automobile producers will be high, as they rationalise and modernise existing plants. Many of the projected investment requirements of major producers remain high up to the end of the 1980s. The requirement for high levels of investment at a time when profits are being squeezed by a slowdown in demand could cause problems for some companies, particularly the smaller companies and those with cash problems, and may also retard investment in flexibile automation.

Flexibility in assembling different models and the subsequent savings from avoiding frequent retooling is an important factor in stimulating the application of industrial robots in automobile production and in reducing, in the long term, the growth of unit costs. However, additional important motives exist primarily aimed at reducing costs. These include:

i) The reliability and accuracy of industrial robots ensures better quality control and consistency in product quality, reducing warranty claims, the need for a large number of quality inspectors (about 10 per cent of an assembly plant labour force), and the rate of rejections, as well as resulting in material saving.

ii) Industrial robots provide the flexibility to cope with temporary peaks in production requirements for various models without the

_____

(Note 16 continued from previous page.)
compete with a world-class manufacturer. In addition, a wide model range assembled in one plant will allow for lower unit costs - such wide ranges are usually available only from the larger producers.

necessity of maintaining idle capacity or large inventories to meet these peaks. Considerable savings through the reduction of inventory requirements are also facilitated through new computerised inventory control systems, robotised warehousing, etc.

iii) computer-aided manufacturing systems increase management control and the ability to reduce bottlenecks;

iv) CAM systems and industrial robots reduce handling time per vehicle by increasing throughput and reducing direct labour costs per vehicle;

v) the reduction in labour costs.

The immediate impact in the introduction of flexible automation will thus be a reduction in, or at least the maintenance of, the real price of new motor vehicles. This is an important consideration, not only in the face of intensified international competition, but also given the need to stimulate new car and replacement demand during the present low growth period.

The present emphasis in the automobile industry on increasing productivity and therefore adopting flexibile automation could have signficant employment repercussions in an industry which in many Member countries is a major employer(17). The impact on labour arises mainly from the intensified international competition which has led to the need to increase productivity. Since labour costs account for about 35 per cent of the total cost of a vehicle, particular emphasis is placed and will be placed on increasing labour productivity and thus reducing labour costs.

The demand situation affecting the automobile industry has led many producers, over the last few years, to reduce their labour force through rationalisation, elimination of second shifts, attrition, etc. It could therefore be argued that the increasing use of flexible automation may result in the maintenance of existing employment levels without significant further reductions in the event of a demand recovery. However, such a scenario is unlikely, given the major emphasis by producers in reducing costs, in rationalising facilities and in modernising plant and equipment. This emphasis is in particular being placed on the increased use of industrial robots in areas

_____

17. For example, it has been estimated that 1 out of every 5 workers in the United States is either directly or indirectly employed in the motor vehicle industries. In France, 15 per cent of manufacturing employment depends on the motor vehicle industry, and in Germany about 14 per cent.

where robotisation techniques are now becoming fairly
standard (e.g. spot welding, painting, handling) and ulti-
mately at assembly areas where in the past cost reductions
have been difficult to implement. In all these cases,
robots are usually used as a direct replacement for labour.

Applications of Robots and their Impact

To assess the impact of industrial robots on the
automobile industry, specific examples of productivity or
labour impact have to be examined. These examples, either
individual company cases or more general situations, are
often of general validity given the fairly homogeneous
production structure in the industry.

Automobile production, because of the variety of
operations, is highly suited to the wide application of
different types of industrial robots and other forms of
automated manufacturing equipment. The production process
for automobiles involves: (i) casting (to produce
cylinder blocks, cylinders, transmission cases, etc.);
(ii) forging (gears, shafts, etc.); (iii) sintering (engine and transmission parts); (iv) heat treating (usually
for hardening parts required to withstand friction); (v)
machine processing and attaching (i.e. machining parts and
assembly of parts); (vi) plastic forming; (vii) metal
stamping; (viii) welding; (ix) painting; (x) assembly, and
(xi) inspection. Many of these processes are already
areas where industrial robots have been introduced suc-
cessfully in the metallurgical industries or within the
automobile industry. In addition, frequent design and
stylistic changes, as well as changes in parts and components, have encouraged the use of computer-aided design
systems and the interface between CAD systems and machining centres(18). The large number of parts involved in
assembly are stimulating use of computer-aided manufacturing systems particularly for inventory control and management.

As the world's largest automobile producer, the production and technology strategy followed by General Motors
will have important implications, not only on the robot
producing industry, but also for other automobile producing companies. GM is presently using about 450 robots

---

18. In the United States, the use of CAD systems in
the automobile industry is growing at an annual rate of
40 per cent. CAD systems are being used extensively, for
example, at Ford Motor Company to undertake all electrical
design work, 50 per cent of new body components and 25 per
cent of other components. At Chrysler, CAD is used to
design 85 per cent of passenger vehicle bodies, 60 per
cent of suspension systems and 40 per cent of engine components. [Source: American Metal Market News,
19th April, 1982.]

and is expected by the end of 1982 to be using, or have on order 1,600 robots. By 1985 the company plans to use about 5,000 robots worldwide, increasing to 14,000 by 1990. Of these about 5,000 will be for assembly use, 4,000 for machine loading and 1,000 will be painting robots. GM's requirements are likely to be upgraded on the basis of further technological advances in industrial robots particularly for assembly use.

The investment cost for GM's programme is expected to be significant since it would include the direct cost of industrial robots (about US$1 billion) plus ancillary equipment as well as plant modifications. Based on a criterion used by GM and other producers that a robot must replace at least two workers to be cost effective, the employment impact would be at minimum a reduction of 28,000 workposts worldwide. The usage and therefore impact of robots will vary from company to company depending on the degree of vertical integration within the company, as well as on other factors. For example, GM has internal stamping, casting and machining plants whereas other companies rely on outside contractors. In contrast to GM's plans, Ford Motor Co. plans to install about 4,000 robots by 1990, less than GM but nevertheless an approximately tenfold increase of Ford's existing stock.

Spot welding lines are presently the main work areas being robotised in the automobile industry. GM and most other producers are aiming at automating about 95-98 per cent of spot welding operations. The spot welding robots relieve human labour of a monotonous and arduous task, furthermore they are reliable and accurate providing consistent quality and higher productivity. The pay-back period for spot welding robots is short - normally two to three years. Most automobile producers increased the utilisation of spot welding robots significantly over the last few years and it is expected that by the mid-1980s most lines will be fully automated. Examples of company use of spot welding robots include Toyota using about 400 robots with another 250 to replace existing welding facilities in 1982. By the end of 1982 most of the companies' spot welding was expected to be automated and by 1983 an additional 200 spot welding robots will complete the automation process. Nissan is presently using about 700 welding robots. As already noted, labour displacement by spot welding robots is normally estimated at two welders replaced per shift so, for example, Chrysler replaced 200 welders over 2 shifts by 50 robots in one plant. At the same time, however, that plant's production rate increased from 50 cars an hour to 65. Fiat, which is increasing its use of robots by 30 per cent in order to increase the percentage of spot welds carried out automatically to 98.5 per cent from 86 per cent has estimated that it will reduce direct labour time per body from 5 hours to 3.5 hours.

Robotised painting booths are another major area of application in automobile assembly. Besides eliminating labour from a health-hazardous area and increasing productivity, considerable savings are achieved in energy costs by eliminating the need for ventilation, and cost savings in paint are estimated at about 15 per cent. GM's painting booths are expected to use 8-10 robots each with an output of 60 cars per hour. Most automobile painting booths can handle several vehicle models and a range of colour changes.

Other examples of robot applications in the automobile industry include:

i) Fiat, which installed robots in 1978 in two automobile plants, boosted production by 15 per cent. The Renault plant at Douai uses 125 robots which are estimated as being 20 per cent more productive than manual labour. Output in the plant was expected to reach 80 cars per man by the end of 1982. Introduction of robots did not lead to labour lay-offs but the labour force is expected to decline over time through attrition.

ii) GM's new Lordstown plant which is presently producing 45-50 car bodies per hour is expected to attain a normal production rate of 75 car bodies per hour. The plant is extensively automated with welding and painting robots. In addition use is made of a robot-operated body inspection system to measure dimensions of J-car bodies in the assembly line. The system is also used to inspect stamped components and subassemblies.

iii) The Fiat-Comau Robogate system provides flexibility by allowing up to 4 different models to be manufactured in the same plant without interrupting the production process. This sytem reduces productions costs since no retooling or equipment modification is required and the cost of investment is spread over more models.

iv) Peugeot has automated its die-casting system for aluminium engine cyclinder blocks and steering column housings with the result that the production rate of engine blocks increased by 50 per cent and 35 per cent for steering gear housings. The automated systems include the use of a robot in the die-casting centre which led to labour saving of 3 workers per shift.

v) BL Cars use a robot for adhesive bonding of trunk lids, replacing 2 workers and increasing

75

throughput by 100 per cent. Robots are also used to apply undersealing to car bodies and are used to test vehicles for watertightness.

vi) The use of flexible manufacturing techniques by Mitsubishi in Japan for the manufacture of its transmission units and engines reduced the required man-hours per unit from 8.8 hours to 2.3 hours and resulted in a 2.5 fold increase in monthly production without an increase in the labour force.

vii) G.M. (Chevrolet) will be bringing on stream in 1983 an automated engine assembly line which reduces labour requirements up to 50 per cent. The plant based on non-synchronous production methods uses robots and other flexible automation.

viii) The Chrysler Corporation is planning on modernising and automating its stamping plants. As part of this programme it is expected that by 1988 nearly 1,000 robots will be introduced, mainly in loading, unloading and transfer operations.

Outlook

The use of industrial robots, computer-aided manufacturing systems and other automated manufacturing equipment is expected to have a substantial impact on capital and labour productivity and on the cost structure of the automobile industry. In terms of labour:

i) there will clearly be a reduction of workposts. Renault have predicted that by 1990 12 per cent of assembly posts in automobile production will have been suppressed. For Volkswagen it has been estimated that 13 per cent of all workplaces can, from a technical point of view, can be suppressed by robots. In the United States, experts have predicted that by 1990 nearly 50 per cent of final assembly for automobiles will be undertaken by programmable machinery. However, existing labour skills will be upgraded from unskilled to skilled labour;

ii) it could be argued that the increased utilisation of flexible automation might lead labour to moderate demands for increases in wages and in non-wage benefits. It is unlikely, however, that such concessions by labour in terms of wage increases would lead to a deferral of investment in flexible automation given that pay back periods for flexible automation are already

76

favourable, and that the expected trend is for
declining machine costs.

    iii) the flexibility offered by robots and CAM sys-
       tems is often more important to producers than
       the labour savings since flexibility allows
       plants to produce profitably even at low levels
       of output.

In terms of capital:

    i) the use of new automated production equipment is
      increasing equipment utilisation, production
      flexibility and easing maintenance of equipment,
      thus increasing capital productivity;

    ii) the increased flexibility in addition to higher
       productivity of flexible automation will, over
       time, reduce investment requirements and facili-
       tate product innovation.

    iii) robots may not necessarily have higher produc-
       tivity than specialised machinery, but their
       flexibility is important in situations where
       expected product variations are high and lot
       quantities variable.

While increased automation is viewed as important by
producers, differences in degree in automation, by different producers or in different countries, are not generally cited as a prime reason for differences in productivity. Thus, for example, it has been estimated that the
assembly time for a subcompact vehicle is 14 worker hours
in Japan compared with 33 worker hours in the United
States. This difference is not attributed to differences
in degree of automation but rather to aspects such as efficient inventory control procedures in Japan (a process
GM and Ford are examining closely), and better management-
labour procedures. The increased adoption of flexible
automation is being undertaken by the European, United
States and Japanese producers so that it cannot be expected that automation per se will narrow international
cost differentials.

The slowdown in demand growth for automobiles has
increased the intensity of international competition and
the drive by producers to maintain existing market
shares. In turn this has led to significant attempts to
improve productivity, improve vehicle performance and
reliability and provide producers with the ability to
rapidly modify design and components to reflect the latest
technological advances. However, while significant employment repercussions in the industry can be expected,
two considerations need to be taken into account. Firstly, in contrast to workposts already being robotised
(welding, painting) which are health hazardous and

monotonous, it can be expected that there will be greater union opposition to the introduction of robots in assembly posts. Secondly, extensive automation in automobile assembly requires a new generation of robots not yet commercially available.

*
* *

6. THE IMPACT OF ROBOTS ON LABOUR

There is common agreement that, in examining the impact on employment of any new technology, the number of factors which affect employment, especially at the macroeconomic level, make it difficult to isolate the particular change attributable to the new technology(19).

In examining the impact of industrial robots on employment, certain characteristics of these capital goods must be recognised. Industrial robots are often a direct replacement for labour employed in manufacturing(20). This is, for example, the case for robots used in welding, surface treatment, handling and assembly. As a result, the direct impact of the diffusion of industrial robots will lead to a reduction in workposts(21).

However, it is also very important that the positive side is stressed. That is, by increasing the productivity of a firm and its competitiveness, industrial robots can play an important role in ensuring the survival of firms and therefore safeguarding employment in those firms. The employment retention effects of flexible automation, both direct and indirect, can be expected to be more important than employment displacement effects in the longer run. Positive indirect effects from the diffusion of industrial robots can also arise from the increase in productivity which improves firm profitability, improves real wages, and improves real incomes through lower product prices and better price/quality ratios. These compensation effects can be sufficiently strong to induce increased expenditure and output in the economy thus counterbalancing direct employment displacement.

---

19. See, for example, Stoneman, P., Blattner, N., Pastré, O., op. cit.
20. There are tasks where robots may be used to replace less flexible capital goods.
21. A reduction of the workforce in the particular firm using robots will depend on the ability of the firm to retrain and transfer labour, the growth characteristics of the firm, etc.

The employment displacement effects of robots must also be compared with the direct creation of jobs (in contrast to saving jobs) as a result of the growth in the robot industry(22). This comparison is unlikely to be favourable, mainly because industrial robots are themselves used in conjunction with NC machine tools and other flexible automation to produce industrial robots. As a result, the production process is highly automated and direct employment creation on the shop floor is low. For example, a case study of an automobile plant in Germany found that, for each industrial robot introduced, the loss of working places was on average 4 workers over two shifts, the number of working places created through maintenance and repair of the robot 0.3 and working places created in production of a robot was 0.5(23). Other studies in Germany have cited the number of workplaces saved per robot as between 0.8 and 6.2, depending on the type of application. The Australian Science and Technology Council reported from a survey of using firms that the average number of workers displaced per shift was 1.3. Estimates from the Netherlands indicate a similar loss of employment.

There is also a need to differentiate between the impact of robots on current employment levels and future prospects. The present acceleration in robot and other flexible automated capital goods installations has not aggravated the existing unemployment situation to any noticeable extent, given that the robot stock is fairly small. For the medium term, it has already been noted that the present trend of applications will continue, changing in the longer term to an emphasis on applications in assembly-intensive industries. Thus it might be expected that the more significant impact on labour resulting from the introduction of industrial robots will occur after the mid-1980s.

An examination of the occupational breakdown of employment in industry can be useful in that, by indicating the relative breakdown of occupations, some magnitudes can be ascertained of the theoretical potential number of workposts where robots might be introduced. While useful, such data must be used carefully since they can provide an erroneous impression that all workposts of a particular type are liable to suppression and since they are also often provided without reference to a time-frame. In the United States, for example, it has been estimated that

----

22. Automation will also lead to a transfer of employment from the enterprise to outside entities (and perhaps outside the manufacturing sector) e.g. to engineering services, marketing, etc.
23. Kalmbach, P. et. al.: Robots' effect on production, work and employment, The Industrial Robot, Vol. 9 No. 1 March 1982.

there are 7 million jobs (or about 50 per cent of manufacturing operatives) where flexible automation, and industrial robots in particular, are of relevance:

Occupation	Number of employees
Assemblers	1,289,000
Inspectors, testers	746,000
Production painters	185,000
Welders	713,000
Packagers	626,000
Machine operatives	2,385,000
Others	1,043,000
Total	6,987,000

Source: Carnegie-Mellon/Booz, Allen & Hamilton "The
Impacts of Robotics on the Workforce and
Workplace"; Quoted in Business Week, 3rd August,
1981.

The above data indicate in particular the large number of assembly workers and thus the potentially greater
impact that assembly robots could have in the future compared to the present generation of robots. The number of
assembly workers is also high in other countries (e.g. in
France the Agence nationale pour le développement de la
production automatisée has estimated that there are
500,000 assembly workers). On the other hand the data
also indicate that a high penetration of robots in painting applications will, for example, have a substantially
lower impact on labour than a similar penetration rate of
welding robots. Similarly a high rate of diffusion of
robots could over the long run have a significant impact
on inspection/testing workposts directly, since robots
have and are being developed for such tasks, and indirectly since higher and consistent quality performance of
robots may reduce the need for inspecting/testing functions.

Some detailed analysis based on expected diffusion
rates of robots and the resultant rates of penetration for
various workposts is shown in Table IV.3. For the United
Kingdom estimated penetration rates by 1990 are fairly
low, in particular for workposts where robot applications
have been proven (spot welding, moulding, painting), but
also for posts such as arc welding and assembly(24).
Similar comments apply to the French projections, where

---

24. It should be noted that the UK estimates were
undertaken early in 1979 and underestimated the significant growth in applications that took place in the early
1980s. For example, by 1981 50 per cent of the stock of
spot welding robots estimated for 1990 were already in
place.

estimates have placed the number of production workers
which could be affected by robots at 300,000. Given the
penetration rates in Table IV.3 these estimates indicate
27,000 redundancies for 1990. The German estimates indicate that the penetration rates in 1985 will be similar to
the UK and French penetration rates for 1990. The German
penetration rates indicate saturation in many workposts.
These estimates indicate that 60 per cent of the maximum
potential posts where robots can be applied will be robotised by the early 1990s resulting in 200,000 redundancies.

Table IV.3

RATE OF PENETRATION OF ROBOTS BY WORKPOST
UNITED KINGDOM
(cumulative to 1990)

Applications	Workplaces Now (Applicable)	Estimated Pene- tration (%)	Estimated No. No. of Robots
Spot welding	36 lines	40	360
Injection Moulding	1,300	50	600
Pressure Diecasting	1,000	50	500
Spraying			
Forging & Extrusion	4,800	10	480
Metalcutting Machines	11,300	5	565
Investment Casting			100
Press Loading	19,000	10	1,900
Heat Treatment	..	..	100
Arc Welding	20,000	10	2,000
Fettling	5,000	5	250
Assembly & Packaging	200,000	1	2,000
Misc. Material			
Handling	200,000	1	2,000
Other	..	..	1,000
Total			12,000

Source: Ingersoll Engineers, Report to the Department of
Industry, UK.

France	1990	Germany	1985	1990
Painting operations	50%	Arc Welding	49%	100%
Metal cutting	50%	Surface Treatment	47%	94%
Gravity Casting	40%	Spot Welding	45%	81%
Spot Welding	40%	Machine Tool	20%	70%
Injection Moulding	20%	Loading/Unloading		
Arc Welding	10%	Die Casting/	22%	54%
		Moulding		
Metal Stamping	10%	Press Work	15%	46%
Assembly	2%	Assembly	6%	40%

Source: Diebold Consultants. Source: Volkholz, V. (op.
cit.)

While estimates of redundancies created in workposts
by the introduction of robots vary widely, it is fairly
clear that the employment displacement effect arising from
the introduction of robots will not be that great. Thus,
on the basis of one possible scenario of the growth in the
robot stock (see Table IV.4), rough estimates by the
Secretariat of the workposts of production operatives
which could be affected(25) by robots indicate that by
1985 about 1.5 per cent of workposts may be involved in
Japan and Sweden, about 0.4 per cent in Germany and the
United States and less than 0.2 per cent in France and the
United Kingdom. By 1990, affected workposts in Japan and
Sweden could be about 3 per cent, 1.5 per cent in Germany,
1.0 per cent in the United States, about 0.5 per cent in
the United Kingdom and less than 0.5 per cent in France.
Such estimates are rough, given that they are based on a
fixed robot/operative displacement ratio and because they
do not take into account labour relocation within firms
which, a priori, could be expected to occur from increased
output levels resulting from higher growth in productivity
and increased competitivity. The impact on labour is also

Table IV.4(1)

SCENARIO FOR GROWTH IN INDUSTRIAL ROBOT STOCK:
SELECTED COUNTRIES
(Units)

	Actual 1981(2)	1985	1990	Growth Rates	
Japan	9,500	27,000	67,000	30% 1981/85;	20% 1985/90
United States	4,500	15,000	56,000	35% 1981/85;	30% 1985/90
Sweden	1,700	4,100	8,300	25% 1981/85;	15% 1985/90
Germany	2,300	8,800	27,000	40% 1981/85;	25% 1985/90
United Kingdom	713	2,700	10,000	40% 1981/85;	30% 1985/90
France(3)	790	2,100	6,500	28% 1981/85;	25% 1985/90

1. This scenario, with the exception of France, is
made on the responsibility of the Secretariat in order to
indicate the potential level of the robot stock, given a
set of growth rates and does not necessarily reflect
Member country forecasts.
2. Source for "actual 1981" data is the British Robot
Association.
3. Source: Diebold FRANCE.

---

25. There is a need to distinguish between operatives
"affected" and job loss per se. For example, workers
could be retrained for new positions or transferred within
a plant without retraining. In areas where applications
of robots are concentrated (welders, production painters,
etc.) a higher percentage of job losses could be expected
among affected workers.

exaggerated since the estimates ignore the dynamic effects of increased productivity on manufacturing in general and any compensation effects arising from increased investment in robots, increases in real wages and price effects which would stimulate demand.

Industrial robots can be expected to result in a restructuring of work skills (a reduction in importance of old skills and increasing importance of new skills) and lead to changes in the work environment and work conditions (shorter hours, less shift work, less overtime). The growth in robotics also has implications for labour training requirements (including white-collar workers, in particular software engineers). The changes in the structure of employment would include (i) the increase in skill levels required for technicians and engineers (ii) the use of less qualified labour in some functions (iii) the obsolescence of some occupational categories. For the most part, robots are expected to improve working conditions, but they do introduce some detrimental aspects in the working environment such as a forced work rhythm, monotony and, to a lesser extent, social isolation.

An important consideration in the impact of robots on the occupational structure of employment is the organisational reponse taken by firms when installing robots. Many firms may prefer a more decentralised production control structure which uses machine operators to maintain and programme robots on the shop floor. Such an organisational response would tend to use and upgrade existing blue collar machine operators. Alternatively greater centralised control of the production process, including the use of computer-integrated manufacturing systems, would involve a greater role for white collar workers in programming, machine setting, etc., and a subsequent reduction in demand for blue collar workers. There is no reason to believe that a centralised organisational response would be the preferred choice of larger firms and a decentralised one the response of medium-sized firms.

On the whole, the response of labour to new forms of flexible automation, and to industrial robots in particular, has been positive in recognising that there is a need to increase productivity and international competitiveness. However, labour representatives have stressed the need for consultation and for open discussion, by firms and governments, on the impact of this new automation on labour. Conversely, labour must be aware that excessive wage demands and non-wage benefits increasing the hourly cost of labour are in themselves important stimulants to the adoption of labour-saving equipment.

*
*   *

7. NATIONAL POLICY APPROACHES

National policies as regards industrial robots and other forms of flexible automation are mainly concerned with facilitating applications, reducing perceived risks in applications, stimulating domestic production capability, and promoting basic research and development. For many countries the scope and emphasis of policy is not yet clear or well-defined given that these countries are in the process of formulating their policy stance (e.g. Australia, Canada) or have just instituted new policies (the Netherlands).

In terms of applications and stimulating diffusion many governments feel that there is a need to make firms aware of the opportunities robotics and new forms of automation provide to improve quality, raise productivity and reduce costs, as well as the scope of robots in improving working conditions. Thus, for example, in the United Kingdom an important aim of government support for robotics is to increase awareness. Consultancies, demonstrations for computer-aided design and manufacturing, management seminars, and an information service are available under a CADCAM Awareness Programme directed mainly at the mechanical engineering industries; similar facilities are available for promoting CAD in the electronics industry. Support for robotics has been approximately £1.3 million per year. Planned expenditure for 1983-84 is expected to be about £4 million.

Although increased awareness per se is an important aspect in stimulating diffusion, the risks inherent in applying a new technology or in installing robots in new types of applications where experience is limited is an important obstacle to overcome. In particular, small and medium enterprises are often faced with hurdles to applications (e.g. lack of engineering knowledge, high initial costs, etc.) not faced by other enterprises. Thus policies for application are also concerned with assisting in the costs of application. For example, the Netherlands is preparing a programme to stimulate application of flexible automation and robotics in industry. Enterprises will be able to obtain government support and low-interest loans for investments in applications of flexible automation that are new for Dutch industry. In addition an advisory group for robots and production-automation (APRA) has been initiated within the Netherlands Organisation for Applied Scientific Research to assist firms with feasibility studies, information and advice.

In Japan a special depreciation programme was implemented (from 1st April, 1980 to 31st March, 1983) which allowed 10 per cent of the purchase price of high performance industrial robots as additional depreciation during the first year in addition to the normal depreciation

allowance. In the United Kingdom selective support for
the industrial application of robotic devices is in the
form of consultancy and/or support for applications (a
grant of up to 33 1/3 per cent of qualifying cost can be
made). In France, CODIS (the Committee for Development of
Strategic Industries) has identified industrial robots
(and automated production systems in general) as a key
area. CODIS basically co-ordinates government financial
support, stressing the need to enhance the international
competitiveness of firms. In 1981 about 50 million francs
were allocated to CODIS and 160 million francs were allocated as assistance to users. The Agence de l'Informatique, created in 1980, is concerned with the diffusion of
information technology. In the United States the Department of Defense (Air Force) has had various projects under
its Integrated Computer Aided Manufacturing (ICAM) project
aimed at encouraging suppliers to use robot systems and
upgrade their manufacturing capability. Similar programmes are being undertaken by the Army and Navy Depart-
ments vis-à-vis their main supplying companies. Canada
has recently implemented a programme of support for tech-
nology-enhanced productivity (STEP) which is specifically
intended to increase productivity, both in the manufacture
and through application of electronics. Forms of assistance include a major projects investment fund (C$ 48 mil-
lion) for the electronics and machinery industries to support R&D and capital manufacturing equipment, including
robots and computer-assisted manufacturing equipment.

In order to assist small and medium companies the
Japan Robot Leasing Company, established in April 1980,
leases industrial robots to these companies. Set up under
the encouragement of MITI, the company is a consortium of
66 firms, including industrial robot makers. Using long-
term prime rate funding from the Japan Development Bank to
purchase robots, the company will lease robots mainly to
small enterprises. During its first year in operation 150
units, valued at Y 1,144 million, were leased to small
enterprises in Japan by JAROL. Within the Japanese context a special financing programme for small and medium
enterprises is provided by the Small Enterprise Loan Corporation in conjunction with the Peoples' Financial Corporation. Under this programme loans, on favourable
terms, are given to small and medium enterprises for industrial safety and health purposes aimed at releasing
workers from health hazardous and dangerous workposts by
substituting machinery. In Sweden, in order to stimulate
the use of modern production engineering, and especially
robots and CAD/CAM, within the process industry and manufacturing industry, SME-oriented development centres will
be established. A condition for the set-up is the active
financial participation of private industry corresponding
to that of the state.

Although for most countries the stimulation of applications is a primary aim of policy, support for promoting

the supply side of the industry is also provided. Such support is often based on the premise that domestic production capability is partly necessary in order to ensure adequate growth in domestic diffusion and successful application. For example, in France importance is placed on stimulating domestic production and in the United Kingdom support for the manufacture of robots is also an important aim of Government support. In the latter case encouragement is provided "for projects involving the design and development by United Kingdom manufacturers of a new industrial robot and associated equipment up to the point of commercial production": a grant of 33 1/3 per cent of qualifying costs is available. In Australia a recent report by the Australian Science and Technology Council recommended measures to facilitate the expansion of robot production but to date no specific government measures have been implemented. The Government has, however, made a reference to the Australian Industries Assistance Commission requesting that it report within a year on the appropriate level and form of assistance for the local robot industry.

An important emphasis of many national policies as regards industrial robots is on research and development. Countries differ somewhat as regards the emphasis of these programmes as well as on the insitutional or organisational aspects. Thus, in Finland assistance to robots is mainly of a research and development nature funded by various government agencies and the Ministry of Trade and Industry. Projects completed over 1975-1979 included automatic, vision-based robot and manipulator systems Further work is under way on automatic, vision-based robot and manipulator systems and research on self-controlling industrial robot systems. In France the Institut national de recherche d'informatique et d'automatique (INRIA) created as a public agency in 1980 undertakes research in information technology and automation with a major focus in 1982 on research on industrial robots. In 1980 the Advanced Automatisation and Robotics Group was founded with participants from research institutes, including the National Scientific Research Centre (CNRS). The aim of the Group over the 1980-84 period was to concentrate on general robotics, advanced remote operations and flexible production systems.

In the United Kingdom the Science Research Council recently launched a £2.5 million research programme on industrial robots. The aim is to examine future generations of robots through a joint academic/industrial partnership programme. Work currently undertaken includes (i) automated assembly of electronic circuit boards and small electronic motors, (ii) free-moving, driverless, warehousing trucks, (iii) robot computer software simulation. An important goal of projects is to ensure that there will be no problems in the introduction of the next generation of robots into the production process. Funding is also

made available to the National Engineering Laboratory and
to the Steel Casting Research and Trade Association (for
studying applications to foundry work). In the Federal
Republic of Germany within the programme for manufacturing
techniques the support area "handling systems" (robots)
includes as major development responsibilities: pattern
recognition, sensor technologies, flexible gripping, drive
systems, programming techniques, flexible peripheral elements, intelligent control units, and safety equipment.
Robot development may also be supported within the programme "humanization of the working process". In Norway
the Royal Norwegian Council for Scientific and Industrial
Research (NTNF), which is responsible to the Ministry of
Industry, has provided grants through 1970-79 for R&D in
robotics and CAD/CAM of about 42 million NOK. During the
years 1980-83 this field will receive about 70 million
NOK, of which 24 million NOK will be provided for robotics
and CAD/CAM for 1983. Norway and the Federal Republic of
Germany have a bilateral joint research project in the
field of CAD/CAM. The Norwegian financial contribution to
this project will be 19 million NOK from 1981 to 1984.

In Sweden the Swedish Board for Technical Development
(STU) responsible to the Ministry of Industry, provides
funding for R&D in robotics and CAD/CAM and is expected to
allocate S.kr.260 million for the period 1980/81-1984/85
for R&D in these areas. In addition there is an agreement
between the STU and the Swedish Association of Mechanical
and Electrical Industries to sponsor a five year (1981-85)
research programme in robotics and CAD/CAM totalling
S.kr.94 million with the share of the STU at S.kr.46 million. An interesting orientation in Swedish R&D is the
programme aimed at automating trouser production. In
Switzerland, research assistance is provided to universities and research institutes particularly to undertake
research on tactile sensors and vision systems, and the
Netherlands has recently initiated a two-phase R&D programme. The first phase (Fl. 2 million) is directed
towards the upgrading of R&D to international standards
carried out on flexible automation and robotics at the
technical universities and at the Netherlands Organisation
for Scientific Research (TNO). The second phase will con-
sist of industry oriented projects.

The major national research and development programme
in Japan, under the auspices of the Ministry of International Trade and Industry includes a 7 to 8 year, Y15 to
20 billion programme beginning 1st April 1983 to develop
remote control robots for dangerous tasks. The proposed
main themes of the project include the creation of a high-
sensitivity sensor capable of object recognition and the
development of a high-versatility robot arm.

In the United States the National Bureau of Standards
and the National Science Foundation support basic research
in robotics and automation. The NSF, for example, is cost

sharing to assist in programmable assembly and through its
Production Research Program provides grants to research
institutes and certain universities to do research on
robots and robot-related work. The National Bureau of
Standards is assisting in the development of advanced
machining systems for small shops, and is also researching
control systems, interface standards and robot sensors.
The US Department of Commerce is considering options on
providing technical assistance and clarification of anti-
trust rules for privately-funded R&D projects. An information network on development in the robotics area is
provided by the Office of Productivity, Technology and
Innovation (Department of Commerce). The NASA also
assists in programmes aimed at flexible automation in
space and maintenance in space.

Chapter V

CONCLUSIONS AND POLICY CONSIDERATIONS

The aim of this report has been to examine the productivity impact of industrial robots on the manufacturing
industries of Member countries. It has been noted in the
Introduction that industrial robots are only one facet,
although a very important one, in the growing trend in the
development and application of flexible automated manufacturing equipment. The interlinkage existing between the
diffusion of robots and other automated equipment must be
stressed in this regard. It may be expected that in the
future this linkage will strengthen, as a result both of
technological developments, and of an increasing tendency
to examine plant production techniques in terms of total,
rather than more partial, automation concepts.

The supply aspects of industrial robots and the be-
haviour and growth characteristics of the robot producing
industry in various countries were examined in Chapter II. The significant high rate of entry of new firms
into the robot industry was noted, particularly of firms
which are expected to have an important role as using
firms, and which have a major multinational role. Differences in the rate of entry of firms as robot producers in
various countries were also noted and this could have implications for the future in terms of relative country
advantages in production and technological ability. On
the other hand the large number of technological, licensing and marketing agreements between firms with an interest in robotics in various countries can be expected to
have positive future benefits internationally.

In examining the diffusion of robots by workpost and
by industry (Chapter III) it has been shown that, by and
large, the pattern of diffusion has been similar in most
major industrial countries. The trend in diffusion has
initially emphasized some important workposts and major
industrial sectors, but the rapid development of technology is expected to lead to wider diffusion by the mid-
1980s to different industries and work areas. In
particular the emphasis on assembly applications will
become predominant. While various factors can be identified which affect demand, no definite conclusion can be

89

drawn at present as to the relative role of the various factors.

The productivity impact of robots has been examined from various perspectives (Chapter IV). Although the present stock of industrial robots is too small to have resulted in any significant impact on manufacturing productivity as a whole, it is, nevertheless, evident that industrial robots can lead to significant productivity increases in the workposts where they are introduced. However, although the productivity increase in particular workposts can be determined on a case by case basis, these increases must be kept in perspective and cannot be extrapolated directly to the firm or industry level. Thus, it is difficult to determine the productivity increases from robotics for a particular plant, for applying industries, or for total manufacturing in general. But, given that robot applications are expected, at least up to the mid-1980s, to be concentrated in a few industries which play a key role in most of the industrialised economies, the positive impact on manufacturing productivity as a whole will be important. For example, in batch-production type industries, which play an important role in many OECD countries, the impact on productivity and cost-savings may be significant in posts where robots are applied. Similarly, in the automobile industry the productivity increases, quality improvements and flexibility in production provided by industrial robots are expected to provide significant benefits in improving the efficiency of this industry. Other important using industries such as the electrical/electronics and machine manufacturing industries will also benefit from robot utilisation. The productivity benefits arising from the introduction of robots can be expected to increase after the mid-1980s as a wider range of industries, and particularly assembly-intensive industries, increase their use of robots.

While the impact on productivity growth of industrial robots is important at the firm level, care must be taken not to exaggerate this impact. The determinants of productivity growth are complex and include management efficiency, and other factors which may constrain the diffusion of the productivity impact arising from an application of new technology. As such the industrial robot must not be viewed as a universal solution to the productivity problem.

It is also important that robots be viewed as part of the overall changes which are taking place in manufacturing concepts with the increasing diffusion of automated manufacturing equipment including computer-aided manufacturing and computer-aided design systems. The impact of new production concepts, equipment and systems on production control and machine utilisation, inventory control, management efficiency will together have a much greater productivity impact than a single machine, such as the

90

industrial robot. In particular, the efficiency of a single tool is not in itself sufficient to increase productivity in a workshop or plant if there is inefficiency in the production flow, or in the other factors of production.

The more general introduction of assembly robots is dependent on the successful outcome of research and development efforts. In some countries, particularly those with a relatively large robot stock, such efforts appear adequate but, for many other countries, there is considerable room for improvement in research, especially where the development of robots needs to conform with the characteristics of the industrial structure. International co-ordination in R&D efforts appear adequate both between firms as well as between various research institutes.

In addition to research and development outlays on industrial robots, many countries have policies to provide an incentive for the production of robots as well as their application. Some countries are mainly concerned with facilitating applications. Support for promoting the supply side of the industry is often based on the premise that domestic production capability is required in order to ensure adequate growth in domestic diffusion and successful applications. There is a danger that the correlation between dependence on imported technology and the level of domestic diffusion may be exaggerated. It may well be more advantageous to disburse government resources on training of qualified personnel (e.g. software engineers), where evident shortages exist in many countries, for use by applying firms rather than in attempting to encourage domestic production.

Given the potential that industrial robots can have in improving productivity, countries may tend to make available to companies investing in such equipment specific incentives and other assistance which are too favourable and go beyond normal investment allowances. Such policies are often contemplated in countries which are behind in the rate of diffusion relative to their major competitors. While this type of measure may be effective in accelerating diffusion of such equipment, selective assistance by governments may often be misplaced since slow diffusion is not the result simply of capital costs. In fact, it is not clear that government assistance is required for robot applications in many areas, given that a fairly rapid pay-back period has been demonstrated. Further, there is a danger that, by providing a capital subsidy, the adoption of such equipment will be stimulated in workposts where labour still has an advantage in terms of cost, thus distorting the optimum balance of factor inputs and accelerating labour displacement. At the international level, problems could also arise if trading partners consider that incentives provided for the development and application of robots are too favourable and unfairly enhance the competitive position of certain products.

The specific case of small and medium enterprises is different, where applications may be retarded by a lack of engineering expertise, easy access to financial resources, and, in some cases, by a reluctance to accept new technologies which themselves are in a process of rapid change. Assistance, particularly on the consulting side, could be useful. There is, however, a need to make all firms aware of the opportunities provided by industrial robots to improve quality, raise productivity, reduce costs and improve work conditions. Governments can play an active role in diffusing this information to firms.

The rate of productivity increase has generally an important impact on the rate of increase in manufacturing costs. More specifically, a high rate of increase of labour productivity will usually imply a lower rate of increase in total cost per unit output vis-à-vis the rate of increase in the hourly cost of labour. In turn, this also implies that a productivity slowdown in manufacturing will have an adverse effect on manufacturing costs as well as in reducing international competitiveness. The application and diffusion of flexible manufacturing techniques, including robotics, by industry can play an important evolutionary role in reducing the rate of increase in costs because of their productivity implications. It is therefore important that Member countries ensure that automated manufacturing equipment can be adopted quickly and without impediments by their manufacturing industries. Policies which facilitate such adoption should be encouraged where there are clear obstacles to adoption.

The rapid growth of robots, CAD/CAM, and other forms of flexible automation over the last few years has reflected the concern of manufacturing industries to increase productivity and thus ensure their future competitiveness. Often these investments have been supported by labour in the realisation that firms with a high rate of growth of productivity and which maintain or increase their competitiveness, also provide the best employment prospects. The failure to introduce and diffuse new automated production techniques in manufacturing industries could have a far greater effect on employment loss, through the loss in international competitiveness, than labour displacement resulting from the widespread adoption of automated manufacturing equipment. The choice is thus between the adoption of such new technology or the acceptance of attrition of an economy's relative competitive position and eventual decline in manufacturing activity.

Since it is not clear that there will be any significant direct labour compensation effects from the production of industrial robots or other flexible automated machinery, the application of such machinery would be facilitated by increased demand for labour in other parts of the economy. The fast diffusion of assembly robots would especially require that the growth environment is favourable, otherwise social impediments could slow down applications considerably.

The need to overcome labour pressure against increasing flexible automation can be facilitated by improved management-labour relations and better information exchanges. In this regard it is important to stress the role of industrial robots in suppressing health-hazardous and dangerous workposts, and in facilitating the upgrading of labour skill requirements on the shop floor. Labour training/retraining policies are therefore of importance.

OECD SALES AGENTS
DÉPOSITAIRES DES PUBLICATIONS DE L'OCDE

ARGENTINA – ARGENTINE
Carlos Hirsch S.R.L., Florida 165, 4° Piso (Galería Guemes)
1333 BUENOS AIRES, Tel. 33.1787.2391 y 30.7122
AUSTRALIA – AUSTRALIE
Australia and New Zealand Book Company Pty, Ltd.,
10 Aquatic Drive, Frenchs Forest, N.S.W. 2086
P.O. Box 459, BROOKVALE, N.S.W. 2100
AUSTRIA – AUTRICHE
OECD Publications and Information Center
4 Simrockstrasse 5300 BONN. Tel. (0228) 21.60.45
Local Agent/Agent local :
Gerold and Co., Graben 31, WIEN 1. Tel. 52.22.35
BELGIUM – BELGIQUE
CCLS – LCLS
19, rue Plantin, 1070 BRUXELLES. Tel. 02.521.04.73
BRAZIL – BRÉSIL
Mestre Jou S.A., Rua Guaipa 518,
Caixa Postal 24090, 05089 SAO PAULO 10. Tel. 261.1920
Rua Senador Dantas 19 s/205-6, RIO DE JANEIRO GB.
Tel. 232.07.32
CANADA
Renouf Publishing Company Limited,
2182 St. Catherine Street West,
MONTRÉAL, Que. H3H 1M7. Tel. (514)937.3519
OTTAWA, Ont. K1P 5A6, 61 Sparks Street
DENMARK – DANEMARK
Munksgaard Export and Subscription Service
35, Nørre Søgade
DK 1370 KØBENHAVN K. Tel. +45.1.12.85.70
FINLAND – FINLANDE
Akateeminen Kirjakauppa
Keskuskatu 1, 00100 HELSINKI 10. Tel. 65.11.22
FRANCE
Bureau des Publications de l'OCDE,
2 rue André-Pascal, 75775 PARIS CEDEX 16. Tel. (1) 524.81.67
Principal correspondant :
13602 AIX-EN-PROVENCE : Librairie de l'Université.
Tel. 26.18.08
GERMANY – ALLEMAGNE
OECD Publications and Information Center
4 Simrockstrasse 5300 BONN Tel. (0228) 21.60.45
GREECE – GRÈCE
Librairie Kauffmann, 28 rue du Stade,
ATHÈNES 132. Tel. 322.21.60
HONG-KONG
Government Information Services,
Publications/Sales Section, Baskerville House,
2/F., 22 Ice House Street
ICELAND – ISLANDE
Snaebjörn Jónsson and Co., h.f.,
Hafnarstraeti 4 and 9, P.O.B. 1131, REYKJAVIK.
Tel. 13133/14281/11936
INDIA – INDE
Oxford Book and Stationery Co. :
NEW DELHI-1, Scindia House. Tel. 45896
CALCUTTA 700016, 17 Park Street. Tel. 240832
INDONESIA – INDONÉSIE
PDIN-LIPI, P.O. Box 3065/JKT., JAKARTA, Tel. 583467
IRELAND – IRLANDE
TDC Publishers – Library Suppliers
12 North Frederick Street, DUBLIN 1 Tel. 744835-749677
ITALY – ITALIE
Libreria Commissionaria Sansoni :
Via Lamarmora 45, 50121 FIRENZE. Tel. 579751/584468
Via Bartolini 29, 20155 MILANO. Tel. 365083
Sub-depositari :
Ugo Tassi
Via A. Farnese 28, 00192 ROMA. Tel. 310590
Editrice e Libreria Herder,
Piazza Montecitorio 120, 00186 ROMA. Tel. 6794628
Costantino Ercolano, Via Generale Orsini 46, 80132 NAPOLI. Tel.
405210
Libreria Hoepli, Via Hoepli 5, 20121 MILANO. Tel. 865446
Libreria Scientifica, Dott. Lucio de Biasio "Aeiou"
Via Meravigli 16, 20123 MILANO Tel. 807679
Libreria Zanichelli
Piazza Galvani 1/A, 40124 Bologna Tel. 237389
Libreria Lattes, Via Garibaldi 3, 10122 TORINO. Tel. 519274
La diffusione delle edizioni OCSE è inoltre assicurata dalle migliori
librerie nelle città più importanti.
JAPAN – JAPON
OECD Publications and Information Center,
Landic Akasaka Bldg., 2-3-4 Akasaka,
Minato-ku, TOKYO 107 Tel. 586.2016
KOREA – CORÉE
Pan Korea Book Corporation,
P.O. Box n° 101 Kwangwhamun, SÉOUL. Tel. 72.7369

LEBANON – LIBAN
Documenta Scientifica/Redico,
Edison Building, Bliss Street, P.O. Box 5641, BEIRUT.
Tel. 354429 – 344425
MALAYSIA – MALAISIE
and/et SINGAPORE - SINGAPOUR
University of Malaya Co-operative Bookshop Ltd.
P.O. Box 1127, Jalan Pantai Baru
KUALA LUMPUR. Tel. 51425, 54058, 54361
THE NETHERLANDS – PAYS-BAS
Staatsuitgeverij
Verzendboekhandel Chr. Plantijnstraat 1
Postbus 20014
2500 EA S-GRAVENHAGE. Tel. nr. 070.789911
Voor bestellingen: Tel. 070.789208
NEW ZEALAND – NOUVELLE-ZÉLANDE
Publications Section,
Government Printing Office Bookshops:
AUCKLAND: Retail Bookshop: 25 Rutland Street,
Mail Orders: 85 Beach Road, Private Bag C.P.O.
HAMILTON: Retail Ward Street,
Mail Orders, P.O. Box 857
WELLINGTON: Retail: Mulgrave Street (Head Office),
Cubacade World Trade Centre
Mail Orders: Private Bag
CHRISTCHURCH: Retail: 159 Hereford Street,
Mail Orders: Private Bag
DUNEDIN: Retail: Princes Street
Mail Order: P.O. Box 1104
NORWAY – NORVÈGE
J.G. TANUM A/S Karl Johansgate 43
P.O. Box 1177 Sentrum OSLO 1. Tel. (02) 80.12.60
PAKISTAN
Mirza Book Agency, 65 Shahrah Quaid-E-Azam, LAHORE 3.
Tel. 66839
PHILIPPINES
National Book Store, Inc.
Library Services Division, P.O. Box 1934, MANILA.
Tel. Nos. 49.43.06 to 09, 40.53.45, 49.45.12
PORTUGAL
Livraria Portugal, Rua do Carmo 70-74,
1117 LISBOA CODEX. Tel. 360582/3
SPAIN – ESPAGNE
Mundi-Prensa Libros, S.A.
Castelló 37, Apartado 1223, MADRID-1. Tel. 275.46.55
Libreria Bosch, Ronda Universidad 11, BARCELONA 7.
Tel. 317.53.08, 317.53.58
SWEDEN – SUÈDE
AB CE Fritzes Kungl Hovbokhandel,
Box 16 356, S 103 27 STH, Regeringsgatan 12,
DS STOCKHOLM. Tel. 08/23.89.00
SWITZERLAND – SUISSE
OECD Publications and Information Center
4 Simrockstrasse 5300 BONN. Tel. (0228) 21.60.45
Local Agents/Agents locaux
Librairie Payot, 6 rue Grenus, 1211 GENÈVE 11. Tel. 022.31.89.50
TAIWAN – FORMOSE
Good Faith Worldwide Int'l Cŏ., Ltd.
9th floor, No. 118, Sec. 2
Chung Hsiao E. Road
TAIPEI. Tel. 391.7396/391.7397
THAILAND – THAILANDE
Suksit Siam Co., Ltd., 1715 Rama IV Rd,
Samyan, BANGKOK 5. Tel. 2511630
TURKEY – TURQUIE
Kültur Yayinlari Is-Türk Ltd. Sti.
Atatürk Bulvari No : 77/B
KIZILAY/ANKARA. Tel. 17 02 66
Dolmabahce Cad. No : 29
BESIKTAS/ISTANBUL. Tel. 60 71 88
UNITED KINGDOM – ROYAUME-UNI
H.M. Stationery Office, P.O.B. 276,
LONDON SW8 5DT. Tel. (01) 622.3316, or
49 High Holborn, LONDON WC1V 6 HB (personal callers)
Branches at: EDINBURGH, BIRMINGHAM, BRISTOL,
MANCHESTER, BELFAST.
UNITED STATES OF AMERICA – ÉTATS-UNIS
OECD Publications and Information Center, Suite 1207,
1750 Pennsylvania Ave., N.W. WASHINGTON, D.C.20006 – 4582
Tel. (202) 724.1857
VENEZUELA
Libreria del Este, Avda. F. Miranda 52, Edificio Galipan,
CARACAS 106. Tel. 32.23.01/33.26.04/31.58.38
YUGOSLAVIA – YOUGOSLAVIE
Jugoslovenska Knjiga, Terazije 27, P.O.B. 36, BEOGRAD.
Tel. 621.992

Les commandes provenant de pays où l'OCDE n'a pas encore désigné de dépositaire peuvent être adressées à :
OCDE, Bureau des Publications, 2, rue André-Pascal, 75775 PARIS CEDEX 16.

Orders and inquiries from countries where sales agents have not yet been appointed may be sent to:
OECD, Publications Office, 2 rue André-Pascal, 75775 PARIS CEDEX 16.

66628-6-1983

OECD PUBLICATIONS, 2, rue André-Pascal, 75775 PARIS CEDEX 16 - No. 42635 1983
PRINTED IN FRANCE
(70 83 02 1) ISBN 92-64-12486-1